*WILLIAM HENRY HOLCOMBE*



# THE SEXES

## HERE AND HEREAFTER

Elibron Classics
www.elibron.com

Elibron Classics series.



ISBN 1-4021-8282-1 (paperback)
ISBN 1-4212-8871-0 (hardcover)

This Elibron Classics Replica Edition is an unabridged facsimile of the edition published in 1870 by J. B. Lippincott & Co., Philadelphia.

# The Sexes.

# THE SEXES

## HERE AND HEREAFTER.

BY

WILLIAM H. HOLCOMBE, M.D.

*Author of "Our Children in Heaven," etc.*

"WHOM GOD HATH JOINED TOGETHER, LET NOT MAN PUT ASUNDER."

PHILADELPHIA

J. B. LIPPINCOTT & CO.

1870.

Lippincott's Press,
Philadelphia.

TO

MY MOTHER AND MY WIFE,

WHO,

ABOVE ALL OTHERS,

*HAVE CONTRIBUTED TO EXALT AND SPIRITUAL-*
*IZE MY CONCEPTIONS*

OF

WOMAN.

# CONTENTS.

## CHAPTER VI.



## CHAPTER VII.



## CHAPTER VIII.

# PREFACE.

---

THE writings of Emanuel Swedenborg, I am confident, are destined to occupy an ever-increasing place in the public regard. Some who are unacquainted with the teachings of this man, suppose them to be near akin to, if not identical with, those of modern Spiritualism. But this is a great mistake. Wide and irreconcilable differences exist between the two systems—if the latter, indeed, may be called a system.

The belief in spirits, both good and bad—in angels and devils—and in the proximity of a spiritual world and the possibility of communicating with it, is warranted by the plain teachings of the Bible, and is the common inheritance of the universal Christian Church. He who denies or doubts these things, has already forfeited his birthright, and is either a rationalist or a skeptic.

Skepticism, Rationalism, and Spiritualism, however, are but protests against the unreasonableness and insufficiency of the old theologies. They are incidental and needful to human progress; the way-stations, but not the abiding-places, of the soul. The New Christian faith has little in common with these phases of opinion. It rises high above them all.

The three central doctrines in Swedenborg's theological system—doctrines repeatedly declared by him to be essential and fundamental—are:

The supreme and absolute Divinity of the Lord Jesus Christ;

The Divinity, and plenary inspiration, of the Sacred Scripture; and

Regeneration through faith in Christ and a life according to his commandments.

These doctrines are seldom met with in the teachings of modern Spiritualism. Nor have Spiritualists, generally, any knowledge of the other distinguishing doctrines of the great Swede—such as those of Influx, Order, Degrees, Correspondence, Heaven and Hell, the Spiritual Sense of the written Word, and the doctrine concerning Conjugial Love.

And although Swedenborg professes to have enjoyed open intercourse with the spiritual world for a period of nearly thirty years, whereby he was enabled to reveal the great facts and laws of that world, yet he repeatedly warns his readers against such intercourse, and shows us why it is attended with danger to men's souls. And those who accept his teachings, generally regard such intercourse as disorderly, against the Divine interdict, and perilous to a man's spiritual welfare.

Upon the subject of the Sexes—here and hereafter—no man has ever thought so profoundly or written so wisely as Swedenborg. He has gone to the bottom of this grand theme. He has made known the spiritual causes of polygamy, concubinage, and prostitution; has revealed the marvelous strivings of the Divine Providence to preserve in man the conjugial principle, and to lead him from a greater to a lesser evil, when it cannot lead him from evil to good; and has unfolded the philosophy and spiritual uses of marriage, and the true relation of the sexes to each other.

And for the benefit of the New Age, whose dawn we already witness—an Age in which the Lord in his Divine Humanity shall reign supreme in the hearts of men—he has revealed a doctrine of Conjugial Love which surpasses in beauty, purity, and practical value, anything hitherto known or imagined

The main purpose of the present work is to popularize some of his teachings on this subject, and to lead thoughtful men and women to give them the consideration they deserve.

"He that hath ears to hear, let him hear."

W. H. H.

New Orleans, La.

# THE SEXES;

## HERE AND HEREAFTER.

---

### CHAPTER I.

*SEX, LOVE AND MARRIAGE UNIVERSAL.*

NATURE," says an eloquent prose writer, "is a System of Nuptials."

"Nature," echoes a charming poet—

> "Nature, with endless being rife,
> Parts each thing into 'him' and 'her;'
> And in the arithmetic of life
> The smallest unit is a pair."

"Loves and marriages," writes Dr. Mason Good, "are common to all nature. They exist between atom and atom, and the philosopher calls them attraction. They exist between congeries and congeries, and the chemist calls them affinity. They exist between iron and the loadstone, and every one denominates them magnetism."

Another acute thinker, Ralph Waldo Emerson, thus formulates the same beautiful truth: "An inevitable dualism bisects nature, so that each thing is a half, and suggests another thing to make it a whole; as, spirit, matter; man, woman; odd, even; subjective, objective; in, out; upper, under; motion, rest; yea, nay. While the world is thus dual, so is every one of its parts."

The grandest generalization of modern science, is thus expressed by Hatch in his Constitution of Man:

"The law of conjugality is the basis of every force in nature."

Sex, love, and marriage, not used in their common and restricted sense, but in their widest and philosophical meanings, are the keys to all the phenomena of mind and matter. Every object in nature is male or female; and one sex is always complementary to the other. Love is the elective affinity, the passional attraction which exists between these complementary forms or sexes. Marriage is the union of the sexes, producing the organization of all things, whether it be the creation of the individual or the perpetuation of the species. Sex is the universal form; love the universal power or force; and marriage the universal result.

Grindon, the most charming of English scientific writers, to whom we are deeply indebted, has thus eloquently expressed the great truth of universal sex and marriage:

"Underlying every phenomenon of the material world and underlying every psychological occurrence, there is found a fixed causative relation of Two things or Two principles, as the case may be; different and unequal, yet of such a difference and such an inequality that, like man and woman—who constitute the type and interpretation of the whole of nature both visible and invisible—each is the complement of the other; one being gifted with energy to act, and the other with equal energy and aptitude to *re*act. All phenomena, alike of matter and of mind, resolve into this dual *virtus*. Whether physical or spiritual, animal or vegetable, Life always presents itself as communicated through one simple formula—*the reciprocal action and reaction of complementaries.*"

"Binary causes lie at the base of all things. The sun and moon cast their light upon us, the rain falls and the waves roll, the spheres preserve their rotundity and persevere in their motions, all as the result of underlying dual forces. The fabric of nature, like its phenomena, resolves everywhere into dualities. Land and water, male and female, the straight line and the curve, do but express prominently a universal principle. The elements, we have already seen, are almost demonstrably only Two."

"In that admirable adaptation and aptitude of things to act and react, and thus to enter into a relation of which marriage is the highest exponent, consists, accordingly, the whole principle of living action. There is no other source of phenomena either in the animated or inanimate world; and wherever it brings things and natures into contact, reciprocally adapted each to the other, life immediately appears, beautiful and exuberant. God made things *complementary* on purpose that they should unite, and open channels wherein his life should have new outlets. Until conjoined and until they have opened such new channels, they are everywhere restless and erratic. Everywhere, in earth and heaven, equilibrium comes of well-assorted marriage or union of complementaries, and there is no equilibrium independent of it."

These advanced truths, foreseen intuitively by a few poets and philosophers, which the present science, psychology and theology can neither appreciate nor utilize, are wrought as foundation-stones into the stupendous system of Emanuel Swedenborg, who has given the world the elements of a new science, a new psychology, a new theology, bearing the same relation to the old which the butterfly, sporting in the golden light of heaven, bears to its first unsightly form in which it groveled in the dust.

All forms are derivations from the uncreated Substance. All causes are emanations from the First Great Cause. Sex, love and marriage are universal principles, derived from the Supreme Being and expressing his nature. The laws of the existence of the Divine Creator are impressed upon and repeated in his creation. Man, the central figure, was created in His own express "image and likeness." "Man is nature concentrated," says Grindon, "and nature is man diffused." Man is the microcosm or little universe, the perfect miniature of nature the macrocosm or the great universe. Both reveal the character and mirror forth the glory of God. Sex, love, and marriage, the fundamental principles underlying both man and nature, must bear upon their faces the image and superscription of the Divine Love and the Divine Wisdom.

"God is both a man and an immortal maid," says an old Greek hymn in one of the Orphic fragments. "Male-female" was also an epithet frequently applied to the Supreme Being by the subtle intuition of Grecian genius. The Divine Love is the feminine, the Divine Wisdom the masculine principle in the Divine nature. They are inseparable, coexistent, co-animating, co-operating. They are the positive and negative poles of the infinite magnet. They exist and are perpetuated by their action and reaction upon each other. As the intensity of each pole of the magnet is determined by the intensity of the other, so the state of the Divine

Love determines the state of the Divine Wisdom, and *vice versâ*. The infinity of one produces and reciprocates the infinity of the other.

The activity of love is Goodness; the form of wisdom is Truth. Divine Goodness and Divine Truth are the sexes of God, yearning for each other with infinite attraction, united together in the divinest marriage. Their reciprocal action is the cause of all life and love and light. The offspring of their marriage is the heavens and the earth and all that is therein.

The magnet illustrates with great beauty the fundamental law of creation. Its male and female elements, known as its positive and negative poles, are concentrated toward its extremities, but cannot be separated. If you break a bar of magnetic iron in two, you will not find one portion positive and the other negative; but each fragment is a perfect magnet as before, with its male and female elements assuming the old relations to each other. The minutest atom of a magnet becomes a magnet itself when separated from its parent one. So is the image of God repeated and perpetuated in every object he strikes off from his own substance, and then reanimates with his breath. Therefore it is that in every dewdrop, every crystal, every organic atom, in every created form, natural or spiritual, the male and female elements stand opposite each other; holding each other in place, inspiring each other with love, impelling each other to use; building up from invisible bases the material and spiritual containants of life.

As the loadstone lying in the bosom of the earth becomes a miniature magnet, receiving power from the great magnet of which the globe itself consists, so does man, receiving his life from God, by a similar induction become a finite form of his Love and Wisdom. Passing into the human being, these forces become the will and the understanding of man—the will being an organ receptive of the Divine Love, and the understanding an organ receptive of the Divine Wisdom. Like the Love and Wisdom of their Divine prototype, they reciprocally yearn for each other in the heavenly marriage; and when the understanding knows and believes the Divine Truth, and the will obeys and loves the Divine Will or Goodness, man becomes an image of God, his son, and an heir to his kingdom.

Every human being, man or woman, is, like the Lord himself, in a certain sense bi-sexual, having both masculine and feminine qualities, which are to be beautifully blended or equilibrated in a spiritual marriage, which is regeneration. This spiritual duality of each individual is represented in the physical duality of the human body. It is composed of two similar halves united at the median line, which are positive and negative or male and female in relation to each other. The entire brain and nervous system, with their wonderful appendages of muscles and bones, are precisely alike on both sides of the body. We have two eyes, two ears, two hands, two feet, two breasts; and where there is apparently only one organ, as the nose or mouth, it is composed of two halves precisely alike and accurately adjusted or married to each other. There are other marriages also in the body; between organ and organ, between function and function, between the nervous fluid and the blood, between the heart and the lungs, between the cerebrum and the cerebellum, etc., too abstruse for popular comprehension; but confirming the exclamation of the Psalmist, that we are fearfully and wonderfully made, and justifying the remark of Galen, that the anatomy of the human body is a sublime hymn in honor of the Deity.

The crowning act of the Divine mercy and glory was the production of two human forms, each external to the other, each the image of God, each a microcosm embodying all the mysteries of nature, and yet with their relative properties and affinities so proportioned and adjusted that each should be the complement, the ideal, the life, the all of the other. Man and woman are bi-sexual in themselves, but in relation to each other they are complementary. In man the masculine element is the positive and the feminine element is the negative principle, while in woman it is exactly the reverse. Therefore it is, by a law common both to spirit and matter, that man is attracted by the feminine element in woman, and woman is attracted by the masculine element in man. From this reversal of the spiritual poles of life, come all the passions and activities of the world; the charms of home, the dream of hope, the purple light of love; the subtlest joys of earth and the eternal bliss of heaven.

Every object in the universe is masculine or feminine. The attractions between these complementary forms, under the generic title of love, produce all the motions and organizations of spirit and matter. The union or marriage of these elements is the vital principle of creation, the secret cause why one thing coheres to another, atom to atom, world to world, and all things to God. It is well for us to contemplate the universality of these truths.

Spirit and matter hold to each other the relation of positive and negative or masculine and feminine. The Spirit of God brooded over the face of the abyss. Spirit is the living, active, impregnating element; matter the passive and receptive. Spiritual forces flow into matter, which gives them form and subsistence, brings them into objective life, as the mother does the babe, and nourishes them from an inexhaustible bosom. Therefore the spiritual and natural worlds are as eternal and as inseparable as the love and wisdom of God. They coexist and correspond, being reciprocally active and re-active; and one without the other is impossible. "The Earth," says Homer, "is the wife of Heaven."

The Sun and the Earth are positive and negative to each other; and it is to their action and reaction that all mundane phenomena are due. The sun impregnates the terrestrial atmospheres with his masculine qualities, and the earth conceives and brings forth all the wonders of the elemental sphere, and all the forms of mineral, vegetable and animal life. When the earth turns herself away from the solar orb, which is called in Scripture the bridegroom, the result is darkness and cold; but when she advances to his ardent embrace, he fires her heart with the warmth and glory of life, and colors her Spring with silver and green and her Autumn with purple and gold.

Heat and Light are the positive and negative solar elements which represent and make operative the Divine Love and the Divine Wisdom in nature. Heat is the feminine principle which expands and opens: Light is the masculine principle, which penetrates and illumines. They co-operate in the creation and preservation of all earthly things. They are the love and wisdom, the affection and thought, the man and woman, of the elemental sphere. Magnetism and Electricity are their counterparts or analogues in a different field, holding to each other a similar relation, and having each in itself a still further combination of positive and negative or masculine and feminine principles.

Land and Water under different forms are repetitions of the same eternal truth. Water is the male or positive element, from whose substance the Land, or female element, was taken or deposited. The Land is wholly negative and unfruitful without the impregnating and fructifying power of the Sea, which draws slowly back from the purple hills all he had given, and rising on the wings of the invisible wind, sinks softly again in innumerable showers upon the bosom of his happy bride.

Instinctively recognizing this truth, man has always called the earth his "mother;" and Homer, the first of poets, expressing the most ancient philosophic opinion, styles water "the father of all things." "Earth," says Lucretius, "impregnated by the liquid rains, produces the luxurious crops and the joyous groves; produces the race of men and all the living tribes." Rivers in all languages are spoken of as males, as Father Tiber, Father Thames, Mississippi—the Father of Waters. The land of Egypt was represented by Isis, and her great river, the Nile, was represented by Osiris, the husband of Isis. And the poetic myth of Venus rising all-beautiful from the sea, appears to be another version of the old story of Eve abstracted from the breast of Adam, or the feminine principle waking into separate consciousness and assuming its objective, complementary and attractive relationship to the masculine.

The masculine and feminine principles are detected in minerals by their relation to the electro-positive or the electro-negative pole of the battery. This binary combination, this essential duality of form and function, this sexuality, so obvious in its simplest expression, pervades the most complex structures, even where it eludes the research of the chemist and is visible only to the clairvoyant eye of the poet and the philosopher.

The sexuality of plants is so apparent that they have been divided into phanerogamic and cryptogamic, or plants with open marriages and plants with concealed or not clearly discoverable marriages. In some forms of the vegetable kingdom the sexes are as distinct to view as in animals; and their interchanges of love are wafted to and fro by the feet and wings of insects and the whispering currents of the concerting winds. When a bee has gathered honey from a male flower, he will alight the next time only on the female flower of the same species, where he shakes from his body the golden dust which impregnates the receptive plant with the aromal life of her distant lover.

The higher the animal in the scale of development, the more complicated is the sexual apparatus, and the nearer to the human type are the passions which its evolution engenders. In the gentle and beautiful birds that bathe their plumage in the light of heaven and fill the forests with their songs, we find the nearest approach to conjugal affection and fidelity. The great difference between the animals and man, proving the immortality of man and the pure materiality of the animal kingdom, is this: that, while the actions and re-actions between the human sexes are independent of time and space, they seem to be governed in animals by periodic and cosmic laws, such as direct the mechanical movements of the tides or the involuntary unfolding of the flowers.

The ancients understood the sexuality of all things better than we do. Keightley furnishes one key to the mythological stories of the Old World, when he says that the Greeks delighted to represent the origin, union and changes of the various parts of nature under the guise of love, matrimony and birth. Causes with them were parents, and effects children. "The division of all mythological beings into masculine and feminine," says Müller, "cannot have been in any event the result of accident." It was rather an intuitive recognition of the universality of a sexual principle pervading all spiritual and natural things, and also of the great truth that the spiritual and natural worlds do not differ in their forms, but only in the substance of which their forms are made.

This great principle of sexuality, flowing from the dual nature of God himself, and repeated in his relation to his Church—for He is the Bridegroom and she the Bride; repeated in the will and the understanding of man, which are to be conjoined in the regenerate marriage; repeated in the forms and forces of the elemental world, in the structure of the vegetable and animal kingdoms, in the happy conjunction of human sexes; repeated also in the relation of Religion to Science, of Church to State, of the sublime to the beautiful in æsthetics, and in a thousand other instances which it would take volumes to elucidate;—this great principle descends from universals to particulars, and every word we speak and every sound we make has a masculine or a feminine quality.

The letters of the alphabet have their sexes. The vowels are feminine and the consonants are masculine elements of speech. The vowels are soft, rounded and fluent, expressing the affections; the consonants are harsh and fixed, limiting the sound and organizing it into an expression or sign of our ideas or thoughts. It is the marriage of vowels and consonants which produces words. Vowels alone produce only sound. Consonants alone cannot be uttered. The consonant and the vowel correspond to the straight and the curved lines of architecture and art, which are the elements of all form and beauty.

Words again are either masculine or feminine. The neuter gender acknowledges only our inability to distinguish the sex of the word. In some words, as in some plants, the sexes are united in the same individual. In the compounding of the words into sentences there are two great classes of words, styled by grammarians the noun-word and the verb-word, which are male and female elements of thought, and without a marriage between which no intelligible statement is possible. The division of composition into prose and poetry still further represents the tendency to assume the male and female forms; for prose, like speech, is the organ of thought, and poetry, like music, is the organ of emotion and affection.

The sound or tone of our voices represents the state of our affections: the words of our speech convey our thoughts or ideas. When the sound and the sense do not strike upon our minds in married harmony, we know intuitively that the hypocrite does not feel or mean what he says. Speech and music also are relatively male and female. Tennyson, intuitively penetrating this mystery, declares that woman must set herself to man

> "Like perfect music unto noble words."

Music is the organization of sound, as speech is the organization of words; and the same laws of sexuality govern in both. Bass is the masculine and soprano is the feminine element in music, which must be married for the production of the highest harmony. The great truth that each human being is bi-sexual, or has both male and female qualities occupying different relative centres in man and woman, is beauti fully illustrated in the distinctions between the musical sounds. Woman's voice divides itself into soprano and contralto, as man's is divided into tenor and bass. Soprano is the voice of woman's affection, contralto the voice of her intellect. Tenor is the voice of man's affection, and bass of his intellect. Contralto is the male voice of woman, and tenor is the female voice of man.

Music is pre-eminently the voice of the affections, charming like woman the most savage breast. It is the subtlest spiritual agency in nature, whereby the very breath of heaven can penetrate into and agitate our inmost souls. "Music," says Mrs. Child, "is the feminine principle, the heart of the universe. What the tone is to the word, what expression is to form, what affection is to thought, what the heart is to the head, what intuition is to argument, what religion is to philosophy, what moral influence is to power, what woman is to man, music is to the universe."

In illustration of the above it may be affirmed that the state of music in any age or country is a fair index of the spirituality or the spiritual potentiality of a people. Barbarians produce nothing but discords. No genuine music is possible among a people generally addicted to polygamy or concubinage. The unrivaled religious music of Europe is due to the chastening influences of Christianity, and to the tender worship of female purity and beauty under the form of the Holy Virgin. And the present advanced and advancing state of musical science, is the best omen of heavenly influences descending upon the world, and the best prediction of the coming reign of woman and of love.

One proof that the creation or created Word and the Bible or the written Word of God have the same Divine Author, is to be found in the fact that the same universal principle of organization runs through them both. Like everything emanating from God and bearing his image, the Bible has male and female elements blended in secret and divine union. Every concrete thought or idea in the Word of God, has a feminine and a masculine side to it, one referring to the emotional, the other to the intellectual sphere of life, whether the subject be the Lord himself, his spiritual church or the individual soul.

This is not the place to prove or even to illustrate this wonderful truth, which appears strange and mystical to those who deny a spiritual sense to the Bible, and who therefore cannot understand the Scriptures nor the power of God. The dual or male-and-female structure of the spiritual sense of the Word, is organic and universal. It has been clearly revealed through Swedenborg for those who have ears to hear. Every incident and verse of the Bible, from the journeyings of the patriarchs and the seemingly trivial ceremonies of the Jewish law, to the prayer-poems of the Psalmist and the dark sayings of the prophet, contain the spiritual wonders of the married will and understanding, and of the vital union between the Divine Bridegroom and his Church.

The Old Testament also contains, relatively to the New, the Wisdom-principle or the masculine and positive element of revelation. The New Testament is the revelation of the feminine or Love-principle. Each Covenant is, again, male and female in itself,—the Law representing the masculine and the Prophets representing the feminine element in the Jewish dispensation; while the Evangelists are the masculine and the Apocalypse the feminine element in the Christian economy. Moses and Aaron were typical in the old law—one of the Divine Truth and the other of the Divine Goodness. Peter and John are the analogous masculine and feminine types in the Church established by Christ. Representing also the heavenly marriage between the True and the Good, our Lord sent out his disciples in couples, "by two and two," to preach the new gospel of peace and love.

The striving of sexual elements through affinities or passional attractions after congenial marriage unions, is the cause of all the motions, growth and activities in the physical and moral worlds. The failure to attain the desired end, and the warfare between uncongenial and repulsive elements, are the causes of all the broken equilibriums, discords and collisions in both spheres. If the atomic marriages in nature were perfect, there would be no storms or droughts or poisons or monstrosities or disease. If the marriage between the individual will and understanding, between the interior and exterior life, were perfect, we should have regenerate men upon earth worthy to be called the sons of God. If the marriage between the sexes were perfect, we should have a social paradise. If the marriage between Church and State, or between our moral and civil ideas, were perfect, we should have a governmental order as sublime and beautiful as that of the stars. If the marriage between the Church and the Lord were perfect, we should have heaven open and the angels of God descending to converse with men.

We have designated the universal power, which draws and binds together the positive and negative elements, by the name of Love. The expression and product of this love is Life, whether it be the gravitative life of atom cohering to atom; the mineral life of oxygen and hydrogen embracing and dissolving into water; the vegetable life of the rose-bush perfuming and crimsoning the air with its presence; animal life displaying itself in happy motions and exquisite songs; or human life bound to the earth by its sensuous affinities and penetrating into heaven with its spiritual aspirations.

Philosophers have long suspected that all the manifold forms of matter are modifications of some simple elemental substance. They are now quite confident that all the so-called forces of nature—heat, light, electricity, magnetism, mechanical motion, etc.—are not distinct as they seem to be, but mutually convertible modifications of one fundamental force. That one force is the Divine Life, appearing as love or heat when it passes into finite, created forms. In one sphere it is the power of gravitation; in another it is chemical affinity; in a third it is electrical or magnetic attraction; in a fourth it is the universe of human feeling and thought. One in principle, protean in form, it assumes a different manifestation for every shape through which it passes. It draws the ocean from his bed; it keeps the moon in her path; it points the needle to the pole; it attracts the flower to the sun; it directs the beast to his prey; it binds man to his home; it leads the Christian to his God.

A brilliant French author gives charming expression to the idea that love is the real divinity that conceals herself, under various chemical and electrical disguises, from the dull eyes of the poor materialist:

"Linnæus said: 'Minerals grow, plants grow and live, animals grow, live and feel'—as many capital errors as words.

"Neither minerals nor flowers are devoid of sensibility; only the sensibility of those inferior beings is not manifested through the same organs as that of man, for the very simple reason that plants and minerals are less richly organized for thinking and speaking than man. But the brain and the larynx are not indispensable in order to feel and to love. Every substance penetrable by electricity, is susceptible of feeling and loving; and all bodies are penetrable by electricity, which plays in nature the part of a universal agent of attraction, of life and of fertility. Electricity operates on all bodies by giving them a sex; that is to say, by doubling them, so as to give to each of their separate parts a strong desire to be rejoined. To love is, properly speaking, to be electrified. It is to feel that one is double, and to feel the necessity of joining the other half of one's being. Man and woman, who are two upon earth, are but one in the other life; wherefore the number of women is equal to that of men in all globes. Electricity preaches by example; and the untiring mutual pursuit of the two sexes is the cause of all the great phenomena of nature."

That love is the secret element or power in universal life, was wonderfully allegorized in the Grecian mythology—that rich treasury of spiritual wisdom. There were two gods known as Love, and still mysteriously regarded as one and the same. These gods were the oldest and youngest of the divine race, standing at the two extremes,—for love is the alpha and the omega, the beginning and the end of all things. The ancient or primeval Love was without parents, or self-existing, and contemporary with Chaos. He was the uniting or blending power of the universe, the cause of attraction, affinity and combination. "On him," as Lord Bacon expresses it, "every exquisite sympathy doth depend." At the other end of the series stands the beautiful and youthful Cupid, the cherubic child of Venus, because beauty gives birth to love, whose special delight it is to enkindle the tender passion. Thus Love stands upon the first and last round of the ladder of development, drawing atom to atom in one place, and heart to heart in another.

Both these gods, so far separated in appearance but identical in reality, were represented as winged infants, blind and naked, and armed with the bow and arrow. These symbols are charmingly significative. Love is represented as a naked infant on account of his tender, simple, beautiful and innocent nature, being the first emanation from the Divine life. He is blind, because it is not his specialty to see, but to feel—for sight belongs to the rational powers or the understanding. He shoots with the arrow, because he radiates his subtle power to a distance like the sun; and he has wings, because it is the aspirations of love only which enable us to soar from earthly to heavenly things.

We have difficulty in recognizing the fact that sex, love and marriage are fundamental and universal, only because we compare the phenomena at the two extremes where the analogies are least apparent; and we see no resemblance whatever between the union of oxygen and hydrogen under the electric spark, and a marriage ceremony before some gorgeous altar, where the loving pair exchange vows and kisses amidst the perfume of orange-blossoms, and the good priest blesses them, as Spenser sweetly describes it, "with his two happy hands." And yet these remote facts are analogous links in the same great chain which binds the least things to the greatest, the circumference to the centre and all to God!

The starting-point is always the same. The first steps are always similar. The embryos of all forms are alike. Yet the universal laws which govern all organization are as active in these simple things as in the most complex. By successive differentiations and evolutions the sexes become varied in structure and function. The simplest unions produce only third substances, differing from both of the constituents. Higher forms of marriage reproduce the species. A still loftier type transmits the spiritual qualities of father to son, so that man becomes "the heir of all the ages." At each ascending series we are delighted with new forms, new faculties, new phenomena; until only scientific analysis can bring us back to the few simple truths—the alphabet of nature—which underlie, pervade, and sustain the splendid universe which has been reared upon them.

Man is organically but one grade above the brute. A single faculty, a single phrenological organ at the summit of the brain, separates him from the animal kingdom. That is the organ of veneration; the idea of God; the capacity to know, to love and to obey a Supreme Being. That spiritualizes his whole nature, moral, intellectual and physical. Veneration makes him negative or submissive to the Divine above him, and sublimely positive to all below him. It makes him immortal.

Everything about the man who reveres and obeys God, becomes spiritualized. His commonest words and deeds are spiritual and are recorded above. His daily life is seen and known and loved by sympathizing angels in heaven, as the islands and ships of the sea are sometimes reflected in the sky. His love is sanctified; his marriage is spiritual. Some sweetly coherent and congenial woman is gradually welded into his very being; and smiling at death, he awakes into life to find his bliss another and yet the same, and all his love that was genuine upon earth eternally renewed in heaven.

We have thus traced the sexual elements, yearning for marriage, from their Divine Source through spirit and matter into elemental and mineral forms, and then upward through all the kingdoms of nature to men and spirits,

> "Until we find, as darkness rolls
> Far off, and fleshly mists dissolve,
> That nuptial contrasts are the poles
> On which the heavenly spheres revolve."

Man sees himself repeated in nature as in a vast and beautiful mirror. Nature indeed is an infinitesimally dissected map of man, revealing his secret structures, both spiritual and natural. This is the æsthetic key to the passion which man displays for investing inanimate things with human feelings and faculties. It is seen best in the innocent sports of young children, and in the works of those wise children who never grow old—the poets. The poets, who, as a quaint writer observes, "half understand God," indulge frequently in the humanization of nature, and secretly charm us by presenting spiritual truths in poetic form, which our feeble and torpid understandings would reject if they were offered as matters of science.

The Latin poet, Claudian, has some lines which Darwin might have chosen as a motto for his poem, "The Loves of the Plants:"

> "The gentle boughs together live in love,
> The happy trees enamored intertwine;
> Palm nods to palm; the poplars softly sigh
> To poplars; alders, poplars, pines and all
> Whisper their tender passions to each other."

How sweetly our great Milton presents us with the same idea, when describing the happy employments of Adam and Eve in Paradise:

> "On to their morning rural work they haste,
> Among sweet dews and flowers; where any row
> Of fruit trees, over-woody, reached too far
> Their pampered boughs, and needed hands to check
> Fruitless embraces; or they led the vine
> To wed her elm. She spoused about him twines
> Her marriageable arms, and with her brings
> Her dower, the adopted clusters, to adorn
> His barren leaves."

The philosopher slowly arrives at his grand induction, that man is nature condensed, and nature is man diffused. The poet sees by a spiritual light that the life of man and nature is one. He detects in all the features of natural scenery the spiritual elements which connect them with the heart of man. He perceives love and gladness and fear and terror in inanimate objects; wisdom in animals; passion in flowers; hymns in the cataract; "sermons in stones;" "books in the running brooks;" epics in the stars; and prophecies in the sunset.

Wordsworth of all the poets has penetrated most deeply into the heart of nature, and evoked her latent sympathies or analogies with man. It was he who turned from the "statelier passions" of his race, to draw from the "humbler urn" of the flowers their "lowlier pleasures;" who forgot his sorrows in "the jocund company" of the daffodils, and his solitude in the presence of the violet; who read his own "unutterable love" in the faces of the clouds, and felt the pulse of his own gladness in the throb of the sunlit ocean; to whom the cuckoo was not a bird, but "an invisible thing,"—"a voice, a mystery, a hope, a love;" and who had so thoroughly read the symbolism of nature and found the secrets of his own heart under it all, that he could exclaim to the little daisy:

> "Sweet, silent creature,
> That breath'st with me in sun and air,
> Do thou, as thou art wont, repair
> My heart with gladness and a share
> Of thy meek nature!"

Swedenborg says that the men of the first or antediluvian church beheld nature as an open book or infinite mirror, in which the good and true things of the Lord's kingdom, descending through their own minds, were outwardly repeated and revealed. In the spiritual world also the surrounding scenery of the angel or spirit is not fixed or accidental, but is a shifting panorama, purely symbolical of his own mental and moral states. The poets only discern, from their high mount of vision, the coming truths which seem mystical to their prosaic fellow-men, but which are really universal and eternal, and the subject of common and daily experience to those happy beings who have escaped our painful limitations of time and space, and have risen into the Land of Light.

Whatever exists in man is repeated in nature; whatever exists in nature may be discovered under some correspondential form in man. Sex, love, and marriage pervade both. The principle of binary combination, caused by the attraction of two complementary forms or forces, underlies every object, every fact, every movement in the world. Sex is the universal form; love the universal force; marriage the universal result. If man be immortal, why are not his sex, his love, his marriage eternal?

## CHAPTER II.

*SEX, LOVE AND MARRIAGE ETERNAL.*

TO him who has discovered the organic bond of connection between the visible and invisible worlds; who can realize the divine solidarity or mutual coherence of all things; who knows that the future life, springing from the same causes and assuming similar forms, is a beautiful, ethereal, and perfected continuation of this; to him theology, retaining its sublimity, has lost its gloom, and death, without ceasing to be solemn, is divested of its terror. To him our last great change, by which we simply disappear from our earthly friends, is not

"So much as even the lifting of a latch;
Only a step into the open air
Out of a tent already luminous
With light that shines through its transparent walls."

When we die, or rather when we seem to die, we sleep; and our sleep is of little longer duration than that which we here experience between one day and another. "This day shalt thou be with me in paradise," said the Saviour to the penitent thief; and there is little doubt that the conversation which was interrupted by the sufferings and death of the cross, was soon resumed between Christ and the new Christian in their spiritual bodies in the spiritual world.

For we awake from the quiet little sleep of death, and find ourselves in a spiritual body.

"There is a natural body and there is a spiritual body," says Paul; which are conjoined during life and separated at death, adds Swedenborg. The soul outside of a spiritual body has no existence; for spirit which has no form has not yet come into being. Man knows himself as "a natural body" whilst in this world, and as "a spiritual body" in the next. The word "sown," used by Paul, refers not, as some readers suppose, to the interment of one's corpse in the grave, but to the birth of our living natural body into the world. "The time," says

John Locke, "that man is in this world, affixed to this earth, is his being 'sown,' and not when being dead, he is put into the grave, as is evident from Paul's own words; for dead things are not sown. Seeds are sown, being alive, and die not until after they are sown."

> "The grave has nothing it can render back:
> We do not pass from nature to the grave;
> But Nature *is* our grave, from which we rise,
> At seeming death, our real resurrection,
> Into the Land of Beauty."

The natural body is subject to disorder, accident, dishonor, and corruption, as Paul affirms. The spiritual body is ethereal, beautiful, incorruptible, immortal. But it is not the less a body, compounded of organs and tissues, and possessed of sensibilities and functions corresponding to those of the natural body. It is also male or female.

It is impossible to conceive of the soul in any other shape than that of the body; and equally impossible to conceive of any other shape which can give expression to the thoughts, feelings, and capacities of the soul. The common perception of mankind, undarkened by metaphysical speculations, teaches that our departed friends are not resolved into spiritual ethers or gases or vapors, but that they exist in veritable human form. Spirits and angels always appeared to the ancient worthies in bodies undistinguishable from those of men. Virgil calls the souls of the dead whom Charon was ferrying over the river Styx, "corpora," or bodies. It is only the transcendental theologian who, by his subtle abstractions, would lead us to the absurd conclusion of Anaximenes, that the soul is a disembodied essence.

The fathers of the Christian Church, very near to the teachings and the traditions of the apostles and of Paul, very generally regarded the human soul as identical with the "spiritual body." Cudworth says: "Tertullian makes the soul itself to be corporeal, figurate and colorate, and after death to have the very same shape which its respective body had before in this life."

Lord Bacon, the founder of the inductive philosophy, echoes this early and rational opinion of the Church. "This spirit whereof we speak," says he, "is not from virtue or energy or act or a trifle, but plainly a body, rare and invisible, notwithstanding circumscribed by space, quantitative, real."

How exquisitely the poet confirms in song the graver teachings of the sage!—

> "Sudden arose
> Ianthe's soul. It stood
> All beautiful in naked purity,
> The perfect semblance of its bodily frame.
> Each stain of earthliness
> Had passed away; it reassumed
> Its native dignity and stood
> Immortal amid ruin!"

Such is the resurrection which awaits us all, even before the hair has ceased to grow upon the dead body our friends have buried. The soul rises in a spiritual body, and that body is male or female. Augustine gravely discusses the question whether women are admitted into heaven in their own bodies, or in the bodies of men, or under some other form; and he comes to the rational conclusion that the female soul must inhabit a female body.

A woman who closes her eyes in this world and opens them in another, has lost nothing of her feminine form and character. She is not a man, nor an hermaphrodite, nor a nondescript, nor a spiritual vapor floating in the mystic ether of universal thought. She is a living, breathing, sensitive woman; and if regenerate, she has every womanly feature beautified and every womanly quality intensified for a higher and better life. The "sex of the soul," as Coleridge calls it, impresses itself ineffaceably both here and hereafter upon the bodily structure. Woman is woman still. Every muscle there as here is a female muscle; every bone is a female bone. From her delicate and *spirituelle* features beam forth the softness and beauty of the feminine soul, and on the elliptic curves of her graceful form accumulates, as in this world, the charming magnetism of life. A man also finds himself organically a man.

The spiritual body is far purer in substance than the natural body; it is like diamond to stone-coal. It has none of those excretory functions which are here necessary to rid us of the poisons we daily absorb. Plastic to the influent forces of the spiritual life, it is a perfect picture and revelation of the soul. It changes with our states of love and wisdom. The good alone are beautiful. The false and evil have their interior ugliness transcribed in baleful characters on their outward forms, while the pure and the loving become more and more perfect and beautiful for ever. In heaven beauty is really what the ancients called it—the outflowering of virtue.

When men and women, newly ascended into the spiritual world, begin to move about and to realize the change which has taken place—to think, examine and be instructed—they discover not only that they have similar yet incorruptible bodies, but also that they have the same mental and moral peculiarities which they exhibited in this life. They have the same sympathies and antipathies, the same faith and opinions, the same appetites and passions, which they had here. They have left nothing behind them but the material body and the natural limitations of time and space. No miraculous transformation has occurred by the act of death. They find that the spiritual life is simply a continuation of this life on higher ground, the evil being totally and for ever separated from the good, and with more potent and heavenly influences and more beautiful and plastic forms, because nearer to the vital source of all being.

Spiritual objects make impressions upon their spiritual bodies precisely similar to those which natural objects make upon our natural bodies. Consequently the life there, although intensely spiritual and intellectual, is very similar to ours. They have an earth as solid to them as ours is to us. They have a sun and a sky and gorgeous clouds that, like ours, are "shepherded by the gentle wind." They have towns and cities and temples and palaces. They walk in gardens and meadows and groves. They have the voices of morning and evening, the glory but not the gloom of mountains, and the green pastures and still waters of enchanted vales.

In that world, as real and substantial as this, love, by whatever name we call it, whatever form it assumes—affinity, attraction, sympathy, passion, aspiration, or adoration—love is the supreme power, and determines the life of the heavenly society and the form of its government. It draws kindred spirits together in the bonds of a thousand shining uses. The overruling love of the Lord and the neighbor blends the affections and thoughts of all souls into eternal harmony, and builds up a heaven within and without. The centre and throne of love's sweetest, highest power in that life, as in this, is the Home. Husband-and-wife, the double-star of the spiritual skies, is the unitary form of heaven. Eden was a type of the celestial country. Every home there is a little paradise in which some happy Adam and Eve, partaking of the tree of life and communing sweetly with God, enjoy each other's love for ever.

Verily, sex, love and marriage are eternal.

This is the necessary and legitimate conclusion to be drawn from the following great and eternal truths: the unity and omnipresence of God; the connection and concordance of the natural and spiritual words; and the immortality of the human soul.

If there be one Creator, indivisible and omnipresent, who operates everywhere and eternally by the same immutable laws, He must have created the spiritual and the natural worlds by the same forces and on the same plan, so that in some manner they may correspond to each other, and be held together by his animating and cementing breath. If sex, love and marriage express forms forces and principles not originating or inherent in dead matter, but imparted to it by spiritual influences flowing from God through the spiritual world, and connecting both spirit and matter with himself, and if man is immortal and retains his identity in a higher world through which the spiritual forces that organized and animated him while here had already passed—the conclusion is irresistible that he will be there just what he was here, minus the material body; that he will be male or female; that he will love and be loved; and that he will be or *may* be eternally conjoined to the object of his love.

Man and woman are so thoroughly and organically male and female, that they cannot be unsexed in the other life without such a radical change in their nature and qualities that they would lose their spiritual identity, and could not, therefore, be said to be immortal. And if not unsexed, then must attraction, affinity, sympathy, passion, and all the forms of love, unless so changed or reconstructed as to be no longer love, manifest the same properties and produce the same effects that they do in this world.

The sexuality of man and woman consists really in the sexual differences between their souls, which are thence anatomically represented in their bodies. It is said in the Scripture that God created man male and female, in his own image and likeness. Surely the image and likeness of God is not in the physical bodies of men and women! That image and likeness must be found, if anywhere, in some male and female principles of the soul itself. Sex therefore is spiritual. If spiritual, it is eternal. Love is the attraction, the very life of the sexes; marriage is their union, their eternal life, their heaven.

This spiritual philosophy of sex, so clear, so beautiful, so consonant with intuition and reason, and having such a sound scientific basis, is quite a new thing to the psychologies and theologies of the present day. The position of the two great ecclesiastical powers on the subject, has been eloquently defined by a writer of great experience in spiritual matters.

"It is obvious that neither Rome nor Protestantism inculcate this faith [marriage in the spiritual world]; nor do they leave it as an august problem to be solved within the portals of a better life. Rome is essentially monastic in its theory. Marriage is a permitted impurity, or at least in holiness far below celibacy. Rome never twined a wreath of celestial flowers for Hymen's brow; never chanted a heavenly epithalamium; never overshadowed the nuptial couch with reverent wings; never diffused a sanctifying power to exorcise the genii who invade and desecrate its mysteries. It leaves the sweetest of all human affections in the grave where the body perishes, and inscribes the sentence of everlasting oblivion upon the mute remains."

"But if Rome is monastic, Protestantism is corporeal. From a celestial point of view its ideas resemble, not winged cherubs, not hymning seraphs, but beasts and creeping things. It marries for time; it divorces for eternity. As a worldly convenience, a temporal morality, a present divine ordinance, it recognizes a mere form of union, and does well when it insists upon its maintenance, denouncing the violations of its ordinances with extreme spiritual penalties. But while it consecrates and ratifies the external bond, it betrays too often the impurity of its internal thought by denouncing as carnal the doctrine that finds in heaven a union of two kindred chastities into one beatitude. The rejection of this faith from its pulpits, the denial of it there, fatally disproves its claim to be considered as the Church of God in any absolute or final sense."

It is necessary to consider the causes of the exclusion of the ideas of sex and marriage from the conception which the Apostolic Church entertains of the life after death. If it can be shown to candid minds that the reasons adduced for such an exclusion are wholly inconclusive, it will go far toward uprooting prejudices and misconceptions which keep out the light of spiritual truth. The doctrines of Swedenborg on this subject, when rightly understood, are found to be rich in heavenly truth and beauty.

The first reason, and a most powerful one, against sex and marriage in heaven, is, that our Lord appears to have settled the question by explicitly declaring that in heaven they neither marry nor are given in marriage. This has precluded all investigation, all discussion of the subject. It has been taken for granted that sexual distinctions are dropped at death, and that we are something entirely different in the other world from what we are here. A separate chapter will be devoted to this important question, and it will be clearly shown that a misinterpretation of our Lord's meaning, necessary and inevitable at that age, has fixed a doctrinal error upon the Church, from which it is about to be delivered by a further unfolding of the Divine Word, and by the progressive development and receptivity of the human mind.

The second reason why the idea of sex and marriage in heaven is repulsive to the orthodox mind, is that the Church has never understood the full significance of marriage as a divine institution and means of regeneration; and that the perversions of the sexual relation have been so terrible, that the common conception of marriage is altogether external, sensuous, and unspiritual.

The world is full of evil, the powers of hell are near to us, and our general range of thought is low and sensual, on account of our corrupt affections and our darkened understandings. It is almost impossible for us to approach the pure and holy subjects of sex and marriage in the proper spirit. Heaven is so closed to us, we live so thoroughly in and for the senses, and our civilization is so cankered with naturalism, that it is difficult for us to see anything in the union of the sexes but a temporary alliance for the reproduction of the species. Its whole world of spiritual uses is closed to us or concealed from us. We suspect there is something unchaste about marriage itself; and our ideal of personal holiness and angelic purity supposes something which is developed apart from and superior to the sexual relation. This idea of impurity connected with marriage, is simply an index of our own impurity.

These false views of the subject have given rise to many absurd and fanatical asceticisms, including the enforced celibacy of the clergy in one Church, and in all of them erroneous conceptions of chastity and of the uses of marriage. To such an insane degree has this phantasy been carried, that at one time the Church regarded woman herself as essentially impure, and she was forbidden to come to the holy communion. Theological delirium could not have engendered a more direful falsity. Akin to it is the delusive idea, that those who separate themselves from the world and devote their lives to holy contemplation and prayer, are especially chaste, especially spiritual, especially accepted of heaven.

Monks and nuns are thought to be married to Christ in a peculiar manner. What a mockery! What a terrible misconception of the heavenly marriage! Every Christian must be married to Christ, or he will never be admitted to the supper of the Great King. Men and women are best married to Christ by being thoroughly, devotedly, spiritually married to each other. They are married to Christ by the faithful performance of household duties, by the care of children, by the love of friends and neighbors and country, by religion in little and familiar things, by subordinating the appetites to the moral sentiments, by honest and useful labors, by mutual trials and temptations and sufferings, by fighting in the world and with the world until they have subdued themselves and the world.

As men can quote the literal sense of Scripture to confirm any absurd or monstrous doctrine whatsoever, they have not failed to denounce the intercourse of the sexes even in wedlock as debasing and unchristian on the strength of the remarkable passage in Revelation xiv. 4: "These are they which were not defiled with women; for they are virgins. These are they which follow the Lamb whithersoever he goeth."

To sustain a preconceived opinion, they have enforced a literal interpretation of this passage, occurring in a book whose every sentence is plainly allegorical and symbolical. Their exposition requires that the mystical number of saints, "a hundred and forty and four thousand,"—representing the whole multitude of the redeemed from the earth, who were not "defiled with women,"—should have been of the male sex. No women at all in heaven! An obvious absurdity! "Women" in the context evidently means the natural affections and passions of the soul; and to be "not defiled with women," is to have kept the heart pure from earthly passions and uncontaminated by evil loves, and therefore "virgin" for the Lamb, who is the Bridegroom of the Church. The passage describes the regenerate soul, and is just as applicable in its spiritual sense to the female as to the male member of Christ's church.

To deliver us from the bondage of materialism and sensualism, a special and extraordinary illumination of the human mind has been granted by the Lord. When we understand the spiritual sense of the Scriptures, which shines through or dissipates the obscurities or "clouds" of the letter; when we see the true relations between spirit and matter, now for the first time clearly revealed; when we are made acquainted with the laws and phenomena of the other life, we shall discover that marriage is the central and pivotal fact of the universe; that the reproduction of the species in this world is the lowest and humblest of its many uses; that the union of the sexes appointed by

God in the first paradise, is the type and image of their eternal union in the last; that wedlock under the divine law, is chastity itself; and that the happy marriage of two loving souls, is so pure, so holy, so immutable, that it is the chosen type in the Divine Word of regeneration and of heaven.

A third potent reason why the world and the Church are not prepared for the grand idea of eternal sex and marriage is, that they know absolutely nothing about the spiritual life, and see no philosophical bond of connection between the visible and the invisible worlds. Religion and science are so thoroughly divorced in the common mind, that men see no special reason for any influx from heaven into earth. They think matter has certain properties inherent in itself, and that material forces and forms have built up this physical world around us, which can run its own course like a machine, separate from and unconnected with any spiritual universe. This latter may be, for aught they know, a very different creation, lying afar off from this, with totally different forms, laws and phenomena, having no necessary and organic connection with this world.

These are monstrous fallacies blinding the mind to any conception of spiritual truth. The first melancholy result is a vicious method of reasoning about spiritual things. Spirit is supposed to be so different from matter, so far from it, so unconnected with it, that they cannot possibly have anything in common. Matter has extension, weight, form, color, odor, organization, etc. Therefore no such properties can be ascribed to spirit. Spirit thus becomes, not something real and substantial, but a mere negation of matter; and as all our thoughts in time and space are based upon our sensuous conceptions, the spiritual world becomes, to the popular mind, simply vacated of everything except certain incomprehensible abstract essences.

What follows? Incapacity to conceive of any real life after death. Practical infidelity as to the existence of a spiritual world. Total indifference, and even intolerance, about spiritual things. It is a melancholy fact that intelligent curiosity in respect to the life after death, is almost dead in orthodox circles. Christians think and know nothing about it outside of the mere question of personal salvation. They are not ashamed to acknowledge their inability and indisposition to form any definite idea of the world they are to live in forever. Their eyes are shut against rational light, or further revelation. They will assent to infinitudes of love, peace, praise, glory, blessedness, etc., in the heavenly kingdom; but they turn a deaf ear when you speak of societies and governments and arts and sciences and churches and homes and husbands and wives in the world to come.

What has reduced the Christian mind to such a vacant and apathetic state? "Do you think we will know our friends in the other life?" is generally the first question to a New Churchman. The apostles, who knew Moses and Elias when talking with the transfigured Jesus, had no doubt about knowing their friends in the other life. The early Christians, who saw devils cast out of the bodies of men, and who met angels at the tomb of the Lord, had no doubt about the reality and proximity of the spiritual world. Has it changed its nature or its relation to this world since that time? John, who received the Apocalypse through one of his brethren, the prophets, had no doubt about the immediate resurrection of the dead, or the opening of the spiritual eyes of a man still living in the world. How the good men from Abraham to Paul, who saw into heaven before death, and recognized there mountains and rivers and animals and trees and tabernacles and temples and sounds and colors and crowds of spiritual beings in human form, would be filled with astonishment at the transcendental theology of the present day, which practically makes the soul an ether, heaven a vacuum, and God, as one great commentator actually defines him, "a luminous abyss!"

Here it is that Swedenborg's rational and philosophical doctrines, illustrating his spiritual theology, render such transcendent service. The spiritual and natural worlds are co-existent, correspondent, and mutually dependent. The forces employed in their creation and maintenance are identical—namely, the Divine Love and the Divine Wisdom operating by the same universal and immutable laws. Therefore the spiritual and natural worlds are not only connected but inseparable; and they are counterparts of each other, each having its sun, its earths, its innumerable forms,—animal, vegetable, and mineral,—and its human beings—for angels are only men transfigured.

All genuine life, force, power, quality, sensation, etc., reside in the spiritual and not in the natural sphere, although the contrary appears to us to be true. What is the consequence? The qualities or properties of matter are not *inherent* but *influent*. Never was a grander truth than this, differentiating the Swedenborgian philosophy from all the feebler systems which have preceded it and prepared the way for it. The ideas of extension, weight, color, form, odor, etc., made on us here, are states of our spiritual organization which exist and are infinitely varied in the spiritual world without the intervention of natural objects. Therefore that world is as real and solid as this—the reality, indeed, of which ours is the shadow.

This is the basis of Swedenborg's doctrine of correspondences, which explains the spiritual sense of the Bible and the spiritual significance of nature. Every natural form is produced by and represents some spiritual form. A pleasant emotion is awakened in the soul; a smile follows in the body, as its representative, or material correspondent. The smile has a spiritual cause and meaning. So has a stone, a flower, a bird, a cloud, the sun—yea, every visible object in nature—its spiritual meaning. Were our understandings sufficiently enlightened and our spiritual perceptions sufficiently quickened, we would catch the spiritual meaning of all things, and the natural world would be to us an open book or mirror repeating and revealing the wonders of the spiritual. Under the old philosophies nature will be always "the open secret" of Goethe; exposed everywhere, explained nowhere; seen of all men, understood by none.

To those who grasp this necessary connection between the spiritual and the natural worlds, together with the priority, superiority, and causativeness of the spiritual, the distinctions of sex after death, the activity of all human passions, and the perpetuity of marriage, are not only rational and proper, but absolutely inevitable. To deny them is to deny the unity and omnipresence of a God working by immutable and universal laws, the relation of cause and effect, the mutual coherence of all things spiritual and natural with one great Centre, and the immortality of the human soul.

The great Puritan poet expresses, in a cautious question, what is the intuitive perception of every mind unbenighted by the ratiocinations of some so-called philosophy:

> "What if earth
> Be but the shadow of heaven, and things therein
> Each to the other like, more than on earth is thought?"

Mrs. Browning, the queen of modern song, recognizes the vital relations between the spiritual and the natural worlds in a more vivid and beautiful manner:

> "Not a natural flower can grow on earth
> Without a flower upon the spiritual side,
> Substantial, archetypal, all aglow
> With blossoming causes."

And as if the veil which conceals the spiritual secrets of nature from obtuser mortals had fallen from her eyes, she exclaims in sublime rhapsody,

> "Earth's crammed with heaven,
> And every common bush afire with God!"

"The characters of nature," says an eminent English divine, "are the hieroglyphics of God." To these hieroglyphics, Swedenborg, illumined from on high, has given us the key; and it will be our own fault if the created and written Words do not enlighten our minds and gladden our hearts with stores of heavenly wisdom hitherto concealed from the world. The richest among these treasures is the spiritual philosophy of marriage on earth and in heaven.

Archbishop Trench, in his work on the Parables, discussing the grounds of analogy between the spiritual and natural worlds, gives unconscious utterance to the Swedenborgian doctrine with great force and beauty:

"It is not merely that these analogies [equivalent to Swedenborg's doctrine of correspondences] assist to make the truth intelligible, or, if intelligible before, present it more vividly to the mind, which is all that some will allow them. *Their power lies deeper than this, in the harmony, unconsciously felt by all men, and by deeper minds continually recognized and perceived, between the natural and spiritual worlds;* so that analogies from the first are felt to be something more than illustrations happily but arbitrarily chosen. They are arguments, and may be alleged as witnesses: *the world of nature being throughout a witness for the world of spirit, proceeding from the same hand, growing out of the same root, and being constituted for that very end.* All lovers of truth readily acknowledge these mysterious harmonies and the force of arguments derived from them. To them *the things on earth are copies of the things in heaven.* They know that the earthly tabernacle is made after the pattern of things seen in the mount.

"It is a great misunderstanding of the matter to think of these as happily, but yet arbitrarily, chosen illustrations. *Rather they belong to one another, the type and the thing typified, by an inward necessity; they were linked together long before by the law of a secret affinity.* It is not a happy accident that has yielded so wondrous an analogy as that of husband and wife, to set forth the mystery of Christ's relation to his Church. There is far more in it than this: *the earthly relation is indeed but a lower form of the heavenly, on which it rests and of which it is the utterance*." Such is precisely Swedenborg's basis for his doctrine of conjugial love.

Archbishop Trench is a distinguished authority on the etymology and significance of words, and he goes on to say:

"Out of a true sense of this secret affinity, has grown our use of the word *likely*. There is a confident expectation in the minds of men of the reappearance in higher spheres of the same laws and revelations which they have recognized in the lower; and thus that which is *like* is also *likely* or probable."

What is more likely or probable than the reappearance in higher spheres of the laws and relations of sex, love, and marriage, seeing they constitute so largely the basis of the physical, social, and historical development of human life?

Thus the analogy or concordance between

spirit and matter, the unity and omnipresence of God, the universality of his laws, their continuous operation in both worlds, the perception that similar causes produce similar effects, the immortality and identity of human spirits, all combine with the inexplicable intuitions of our own hearts to assure us that the life to come is essentially a continuation of this on a subtler and grander field, having similar forms, affections, and relations both personal and social.

Let no one suppose, however, that there will be no difference between heaven and earth. Earth at its best is but the shadow of heaven, which immeasurably exceeds it in the perfection and glory of its forms and the rapidity and beauty of their changes. Freed for ever from the disorganizing influences of evil, lifted into purer and sweeter atmospheres of life, the affections and thoughts of the soul will expand indefinitely. The peace and joy of personal relations in the Hereafter, the harmony and strength of social institutions, transcend our present greatest conceptions. Our love and knowledge of God will be perpetually increasing, and He will be thereby enabled to give us ever new and more glorious revelations of his goodness and power. Eye hath not seen, nor ear heard, more than a poor fraction of the heavenly life as it is, and no heart hath imagined its possibilities in the future.

Oh the hopes, the fears, the longings, the imperishable activities of conjugial love! Its intensity reminds us of what Paul said about the love of Christ. Indeed, the marriage of Christ with the Christian heart is typified in the inseparable union of two congenial souls; and the fervid words of the apostle are as applicable to the true husband and wife as to the Bridegroom and his Church:

"Who shall separate us from the love of Christ? Shall tribulation, or distress, or persecution, or famine, or nakedness, or peril, or the sword?

"Nay, in all these things we are more than conquerors, through Him that loved us.

"For I am persuaded that neither death, nor life, nor angels, nor principalities, nor powers, nor things present, nor things to come, nor height, nor depth, nor any other creature, shall be able to separate us from the love of God which is in Jesus Christ our Lord."

Like Solomon's song, this is the language of conjugial love as well as of religious fervor.

How sweetly does that pure-hearted and saintly woman, Mrs. John Fletcher, plead for a perfect reunion with her great and good husband, who preceded her thirty years into the kingdom of light and love!—

"As spiritual union arises from a communication of the love which flows from the heart of Christ, I cannot but believe that a nearer approach to its centre, and a fuller measure of that divine principle, must increase and not diminish the union between kindred souls. Will not, it is asked, all particular unions cease? and is it not the design of God that death should divide husband and wife? To answer this objection I must premise that, *what is of God shall stand.* I plead only for that union which has God as its source; and I think it will not be hard to prove that, *what God hath joined together, death cannot put asunder.* Division comes not from God, but the devil."

Far away—the moral antipode to this pure

and saintly soul—we find the gross and reckless Byron, in some happy moment of spiritual insight, felicitously expressing the same truth:

> "There are two souls of equal flow,
> Whose gentle streams so calmly run,
> That when they part—they part? Oh no!
> They cannot part! those souls are one!"

Indeed, the poets are all with Swedenborg on this question. For the perception of spiritual truth, they are more trustworthy than theologians or philosophers. They stoutly ignore the fact of death, and color time and eternity with the purple splendors of love. The expounders, the custodians, the very artists of love, they trample boldly on theological and popular fallacies, and speak of the reunion of husbands and wives in heaven, as if it was a matter of course and to be expected by every rational man.

One of the grandest passages in Tennyson or any other poet, is where King Arthur takes his farewell leave of the guilty Guinever, who had wrought the ruin of his kingdom and his home, and who lay groveling at his feet upon the floor. The stern reproaches of blighted glory and honor softly subside into expressions of forgiveness and pity, and the peerless husband, heartbroken and marching to his doom, begs of the penitent wife, from whom he felt himself inseparable, the hope at least of her purified love in heaven:

> "My love through flesh hath wrought into my life
> So far, my doom is that I love thee still.
> Let no man dream but that I love thee still.
> Perchance, and so thou purify thy soul,
> And so thou lean on our fair Father, Christ,
> Hereafter, in that world where all are pure,
> We two may meet before high God, and thou
> Wilt spring to me and claim me thine, and know
> I am thy husband—not a smaller soul,
> Nor Lancelot, nor another. Leave me that,
> I charge thee, my last hope."

And the remorseful queen, who dedicated herself to humility, chastity, and devotion, until she passed purified

> "To where, beyond these voices, there is peace,"

treasured this little hope bequeathed her by the king, as the sacred germ of the new life springing up in her soul.

"What hope? I think there was a hope,
Except he mocked me when he spake of hope;
His hope he called it; but he never mocks,
For mockery is the fume of little hearts;
And blessed be the king who hath forgiven
My wickedness to him, and left me hope
That in mine own heart I can live down sin,
And be his mate hereafter in the heavens
Before high God."

There is a beautiful Grecian myth versified by Bulwer in his "Lost Tales of Miletus," which not only teaches marriage in heaven, but the still more advanced truth that those marriages are determined by the inmost, ineradicable, indestructible spiritual affinity of the two souls, independently of anything which occurred during their earthly lives.

The argument is this: Leonymus of Croton, invading Locria, was wounded by a phantom figure of Ajax, "guarding still his native soil." This ghostly wound could not be healed by any physical means; and the sufferer was sent by the oracle to the Isle of Happy Souls, where he was told he could be cured by the touch of the spirit-hand which had inflicted the injury. The poet does not tell us how Leonymus got into the spiritual regions in search of his homœopathic cure; but, when he did get there, he found a beautiful external nature remarkably like ours, and Achilles, the greatest warrior of Greece, and Helen, its most beautiful woman, enemies in the earth-life, enjoying an eternity of wedded bliss.

The visions of the prophet and the intuitions of the poet have been verified by the observations of Swedenborg, who teaches, not from theory or fancy, but "from things heard and seen," as he himself expresses it, that men and women are still male and female in the other life; that they are drawn together there by the same spiritual affinity which attracted them here; and that in heaven the relation of husband and wife is established on a sure, beautiful, holy, and immutable basis.

How beautiful, how real, how near the other life becomes, when we view it from this standpoint! All that has been said or sung of the charms and pleasures of home, and of the unspeakable delights of newly-wedded love, is "imaged there in happier beauty." The perfected married pair is the type of the Lord's

Church, the miniature form of Heaven, the centre of divine influx, the repository of divine life, the crowning work of the divine wisdom, the consummated flower of the divine love.

# CHAPTER III.

*WHAT OUR LORD SAYS ABOUT IT.*

IN the resurrection they neither marry nor are given in marriage; but are as the angels of God in heaven."

So spake Jesus to the Sadducees, who told him of the woman who had seven successive husbands, and who asked him, After death, whose wife shall she be?

Marriage was established by the Lord himself in the garden of Eden. It was the first and holiest institutional work of his hands; preceding his Church, preceding his Word, preceding his Heaven,—for all angels have been men, and are the offspring of marriage. Before the fall, man and woman were husband and wife in the purest innocence and love; and they were then, what they have never been since, the perfect image and likeness of God, precisely as the angels are now.

The Lord has perpetuated and blessed this heavenly institution which He then ordained. "Houses and riches," says Solomon, "are the inheritance of fathers; but a prudent wife is from the Lord." "Marriage is honorable in all, and the bed undefiled," says the apostle.

Marriage is used as the type of the union of the Lord with his Church; and the wanderings of the heart from God, and the consequent falsifications of his holy Word, are described as conjugal infidelities which leave their darkening and deadening effects upon the soul. Marriage is also the type of regeneration and of the kingdom of God in the individual spirit,—of the union of God's goodness and truth in the individual life. How holy must be the *essence* of marriage when such is its signification!

Our Lord's first miracle was performed at a marriage-feast, which represented that union of goodness and truth, of charity and faith, in which alone the divine power and presence become manifest. The water was turned into wine, to represent how the natural truths of the Word are made spiritual to our perceptions by a life according to the commandments. By the same process our earthly marriage is converted into a heavenly one.

Marriage is the great civilizer of man; the organizer of society; the peace-giver and joy-giver of the world. Its condition among a people is the true measure of their spirituality. If all men and women were married and their marriages were perfect, wars would cease; diseases would disappear; supreme order would prevail; love universal would reign; heaven would descend to earth.

If such be the origin, nature, and effects of marriage, why should it cease with this life? If sex is spiritual and love is spiritual and marriage is spiritual, why should they not also be eternal? Who is to divorce the two souls that have been wrought into one by the unitizing operation of God's own law? Are there recognitions and reunions in the Happy Land for parents and children, for brothers and sisters, for friends and neighbors? and shall they whose souls and bodies have been made one by divine institution, whose beings are knit together by the sweetest, holiest, profoundest sympathies,—shall they meet each other merely as neighbors and know each other merely as friends? Impossible! Why should the conjugial love alone perish?—conjugial love, the fountain of all other loves, and itself the first and sweetest gift of God!

Nothing denies or questions the eternity of marriage; nothing in the laws of nature, in the structure of the soul, in the analogy between the visible and the invisible; nothing in the light of reason, in the intuitions of the spirit, in the revelations of God; nothing whatever but the single sentence, "In the resurrection they neither marry nor are given in marriage."

Let us examine this solitary utterance, which stands as a seeming discord in the universal harmony. Let us approach it with reverential faith. It is no mistranslation. It is no interpolation. It is an utterance of the Divine Truth. Heaven and earth may pass away, but not the Word of God; neither the whole nor a part—not one jot or tittle of it. This sentence may not teach what we think it does; but whatever it really teaches is supremely true; and all our theories, doctrines, systems, and interpretations, must fall to the ground before it.

We must not be so hasty and self-confident as to suppose that we see the Lord's meaning at a glance; that the truth shines on the surface; that the sense most obvious to every reader is the genuine sense. Natural things may sometimes be seen at a glance, because they are discovered by the natural senses; they shine on the surface because they are superficial. Spiritual truths are all interior and coherent, woven together by spiritual laws, unseen by the carnal eye, and symbolized in Scripture by forms and modes of speech which need interpretation.

"Whoso eateth my flesh and drinketh my blood hath eternal life," said Jesus in the language of spiritual symbolism. The Jews "strove among themselves, saying, How can this man give us his flesh to eat?" His bewildered disciples also said, "This is a hard saying; who can hear it?" And the Saviour left the Jews in their mental darkness; but he declared to his disciples that his intention was to convey spiritual truth under natural or literal forms of expression. "Doth this offend you?" he inquired. "It is the Spirit that quickeneth: the flesh profiteth nothing: the words that I speak unto you, they are spirit and they are life." The explanation was insufficient. Many of his disciples went back and walked no more with him; and he said sorrowfully to the twelve: "Will ye also go away?" So hard is it for the unregenerate mind to recognize spiritual truth under its corresponding literal forms, to which it insists upon giving nothing but the most obvious, literal interpretation.

"The letter killeth," says Paul, "but the spirit giveth life."

The spirit is needed to give life to many parts of the Scripture which the letter has killed. The first eleven chapters of Genesis, evidently so different from all the rest, the singular incidents of Jewish history, the apparently trifling details of the Jewish law, the dark sayings and unintelligible symbols of the prophets, and the mystical visions of the Apocalypse, are all instinct with spiritual life and beauty, needing the voice of the interpreter to decypher their hieroglyphics and to utilize their transcendent truths. Happy will it be for the Church and the world when they recognize the great fact that the Second Coming of the Lord is not a personal appearance in the clouds of heaven, but a new revelation of the Divine Wisdom through the clouds—or literal sense—of the Sacred Scripture.

There is, then, a spiritual sense to the Bible. The true meaning of any passage of Scripture is always found in the spiritual sense. Why should the literal and spiritual senses ever seem to differ? Because when the natural mind is incapable of receiving spiritual truth in its purity, it always gives it a natural or literal interpretation. This is not the place to discuss thoroughly the coexistence of two senses in the Bible. We affirm the fact, and say that in the most obscure and difficult passages we find brilliant and glorious truths in the spiritual sense, which do not in the least appear in the letter.

"All these things spake Jesus unto the multitude in parables, and without a parable spake he not unto them.

"And when they were alone, he expounded all things unto his disciples.

"And the disciples came and said unto him, Why speakest thou unto them in parables?

"He answered and said unto them: Because it is given unto you to know the mysteries of the kingdom of heaven, but to them it is not given."

It is plain, therefore, that the words of Christ contain a double meaning, an internal and an external. The external sense fell upon the ear of the multitude in the shape of parables or mystical sayings, to which each man gave a literal interpretation according to his degree of natural light, or his own analytic and imaginative powers. The other sense, explaining the parables and unfolding the "mysteries of heaven," was reserved for those who could retire from the world "alone" with Jesus, and become receptive of truth in a higher or spiritual degree. The same distinction exists to-day. The literalists stand outside with the multitude, and are not permitted to know "the mysteries of heaven" which lie concealed in the internal sense of the Word.

Thus also one dispensation of truth is interior to and higher than another, not to be understood by those who occupy the lower and more external standpoint. Paul told the Jews that the story of Hagar and Ishmael was an allegory involving spiritual truths which they had never suspected to exist in it; and that the veil which Moses had put over his face, concealing the divine light, was still upon their own hearts when Moses was read in the temple. Swedenborg has been empowered to reveal the still deeper meaning hidden in the Scriptures, and to lift the veil of both Moses and Paul from the mind of the Apostolic Church.

We are not wandering at all from the subject of marriage in heaven, as the reader will soon discover.

As the soul or living spirit requires a natural body for its outward manifestation, so the living spirit of God's Word required a literal form or body for its outward expression. As our natural body is feeble and imperfect from its necessary limitations in time and space, and is thus a poor and inadequate manifestation of the immortal soul within, so the letter of the Word, being expressed through the finite minds of a feeble and sensuous race, takes on their imperfections, and reveals in its external forms very little indeed of the divine glory which illuminates its interior. To insist that the divine revelation lies in the letter of the Word alone, is like attributing all life and power to the material body, denying the spiritual element within it.

Without the body, no soul could be manifested. Therefore the letter of the Word should be the object of scrupulous care. On its preservation, its accuracy, its purity, depend our hopes of getting at the genuine truth within. But to limit the revelation of divine truth to the letter alone, is to circumscribe, depreciate, and really dishonor the gospel. It is the demand of the sensuous skeptic that a revelation from God shall be perfectly simple and intelligible to all, bearing its meaning on the surface, and as plain and clear as the edicts of an earthly king. And this were a reasonable demand if the real spirit and meaning of the Word is to be drawn from the letter. But the skeptic finds the Bible full of strange and miraculous histories, mysterious prophecies, and inexplicable visions, about the true meaning of which the whole Christian world is sadly divided in opinion; and he rejects it all as the work of imposture or fanaticism.

The Christian who is capable of spiritual thought above the senses, who believes the Bible to be inspired, and to contain the wisdom of Him who created the universe both of mind and matter,—he ought to be the first to acknowledge that the Scriptures may, and indeed must, have meaning within meaning for angels as well as men; that, being constructed by the Author of the visible and invisible creation, and according to the same immutable laws, the Bible must be as vast, as deep, as complex, as myriad-sided as creation itself. Its essential divinity cannot, therefore, reside in the mere letter.

It is a great error to suppose that spiritual interpretation would give a loose rein to the imagination, and flood the Church with wild theories and fantastic speculations. Nothing could be more remote from the truth than this. It is the bare letter which has done all this harm. It is the letter which has torn the Church with schism and filled it with false doctrines. It is the letter to which every sect appeals in support of its peculiar tenets. It is the letter which has created the carping spirit of rationalism, and winged the shafts of infidelity. How strange that the Christian Church should not yet have generally discovered that "the letter killeth;" that "the flesh profiteth nothing;" that the Lord's words are spirit and life; that the Holy Book is written *within* as well as *without;* that the rock conceals water in its bosom; that the water contains wine interiorly; and that the inmost of the wine is the blood—the living spirit—of the Lord!

The spiritual sense revealed by Swedenborg bears no resemblance to the fanciful traditions of the Hebrew Cabalists, and scarcely any to the ingenious speculations of Origen and other Christian writers. It was not imagined, conceived, discovered, or invented by Swedenborg at all. It was revealed to him from heaven by a special illumination of his mind. It has always existed in heaven; it is the food of angelic minds; and its revelation to men will reconnect heaven and earth. This spiritual sense is indivisible like the inner garment of the Lord. It is a fixed, organic, coherent, unitizing system of Divine Truth, which cannot, like the letter, be interpreted in different ways.

Convinced, now, that there is an interior,

higher, and spiritual meaning to what our Lord says, which may or may not shine through the outer and literal statement, let us consider his remark about marriage in heaven. Remember that he always means spiritual things, when the literal expression appears to relate only to natural things. The water of which he spake to the woman at the well, was spiritual water; the meat which his disciples knew nothing of, was spiritual meat; the wine and bread of the holy supper were spiritual elements—his own goodness and truth. When he spake of death, he meant spiritual death; of the resurrection, he meant a spiritual resurrection; for he says "I am the resurrection." When he alluded to father, mother, and brethren, he meant spiritual relationships, disclaiming all others. When he spoke of marriage, however gross and carnal *our* idea of it may be, he meant spiritual marriage, *and that only.*

Spiritual marriage? What is that? There is a common opinion that although there is no marriage in heaven in the least resembling ours, still there is a mystical union of kindred souls—a blending together in eternal harmony of tender thoughts and gentle sympathies, so that two lives may become spiritually one. Swedenborg has dissipated the obscuring mists which have hitherto surrounded this subject, and by actual observation and experience has revealed the constitution of the spiritual world, as the scientific men of our earth have discovered the constitution of the natural world.

He finds that the spiritual and the natural worlds correspond to each other. He finds that everything which exists in the natural world is repeated in the spiritual world in spiritual forms. We have shown in our first chapter that everything in nature is bi-sexual; that electric and magnetic polarities divide every object and every atom in every object into positive and negative or male and female forms, which have for each other special affinities, and are forever striving for equilibrium, union, or, as we prefer to say, marriage.

This great natural law is an image or correspondence of the spiritual law, that every spiritual force or form is positive or negative, male or female. Every idea of the mind is compounded of the affections and thoughts which give it outward expression. When the affection (female) is married to the thought (male) we have the outward manifestations of life, sensation, perception, action. Forms or objects are bound together in the spiritual world by affinities similar to those which exist in the natural world. The female forms have relation to love or goodness; the male forms have relation to truth or wisdom. Their marriage produces the outer world, whether spiritual or natural. Goodness alone, or truth alone, is nothing; combined, they are all things. They only become objective when subjectively united.

This great truth, lying at the basis of all cretions, is thus beautifully expressed by a heaven-inspired poet:

> "Such perfect friends are Truth and Love,
> That neither lives where both are not."

The first or primary form of marriage is the perfect union of the will and understanding, of the affections and thoughts, of the emotions and the intellect, of charity and faith, in the individual soul. The regenerate man is the man in whom pure and holy affections are so wedded to true and wise thoughts, that his entire life is divinely moved to good actions.

A more complex form of the same marriage exists between the sexes, both here and hereafter. Man and woman alike have in themselves individually the elements of goodness and truth, and must each attain that marriage of pure affection and wise thought, which is regeneration. But sexually they are different. Woman is the form of goodness, man the form of truth. These exercise the same love or attraction for each other as the affections and thoughts do in the individual mind. Men and women love each other just as each individual loves what he believes and believes what he loves.

A still more complex form of marriage is between the Lord and his Church. Here the married couple, regenerate in themselves, regenerate in their union with each other, make the unit or least form of the Church, which is female or negative in relation to the male or positive Divine Humanity of Jesus Christ, which animates and impregnates the two-souls-in-one with all the glories, beauties, harmonies, peace, order, love, of the spiritual and eternal life. To one of these three marriages,—the marriage of the will and the understanding in the regenerate life of each individual; the marriage of the Divine Goodness and the Divine Truth represented in objective male and female forms, husband and wife; or the marriage of a spiritual being compounded of two congenial beings to the great Bridegroom of the Church,—to one of these three marriages, and indeed to all of them in turn, every scriptural sentence alluding to marriage, when rightly understood, will be found to refer.

The reader may hastily conclude from this explanation, that our Lord, in saying that "in the resurrection they neither marry nor are given in marriage," has affirmed that in the spiritual life there is no union of the good and true, no marriage of charity and faith, none of the very things which we have described as the essential and organic elements of the spiritual world. A closer inspection, however, will convince him of his error.

The interpretation of the passage evidently turns upon the meaning of "marrying and giving in marriage." About the natural signification of the words there can be no doubt. But most readers have never yet learned the fact, that the spiritual signification of words and things is twofold. The same words in the Bible have two different and opposite senses, according as they refer to good or evil, to the true or the false, or according as they relate to heaven or hell.

The fundamental cause of this is, that the evil and false have no absolute and independent existence, but are simply perversions of the good and true; as disease is not a real object or power, but only a perversion or disturbance of the natural healthy functions. The good and the evil spiritual powers are often symbolized in nature under the same forms. Thus there is a marriage which is not only possible in heaven, but which is heaven itself; and there is also a marriage not only impossible but incomprehensible to angelic beings. And both of these spiritual marriages are represented by the natural symbol—marriage—as it comes to our natural perception.

This is a point of so much importance to an

intelligent comprehension of Scripture mysteries, and of this one in particular, that we must press its careful consideration on the reader.

Let us take the word *water*, and we shall see that it has two opposite significations according to the context.

The waters which gushed from the rock in Horeb, to save the dying multitudes; the "still waters" associated with the "green pastures" in the land of the Good Shepherd; the water of life, "clear as crystal;" and the consecrated water of baptism, are all evidently significative of divine truth in its regenerating influence on the soul. But when the Psalmist exclaims, "Save me, O God! for the waters are come in unto my soul," he evidently means some kind of waters very different from the preceding. He evidently means no natural waters whatever. He means those false and deadly persuasions which would assault and destroy his faith in God. He means falsities opposite to the truths of heaven; for water in its good sense refers to truth, and in its opposite sense to falsity.

Take the symbolic meaning of the word

*sword.* "They that take the sword shall perish by the sword" is the solemn warning of the gospel. "They shall beat their swords into ploughshares," is a prediction of a state of heavenly peace. But the same gospel, in apparent though not real contradiction to itself, says, "Prepare war: beat your ploughshares into swords;" and again: "He that hath no sword, let him sell his garment and buy one." These are respectively the swords of truth and of its opposite falsity—the one used in the defence of the good, the other in the aggressions of the evil; and the meaning demanded in each place is determined, not by the word, which is twofold in its signification, but always by the context.

The Bible recognizes two meanings to the word *peace.* How tenderly does the great Prince of Peace touch the hearts of his disciples by those sweet words: "Peace I leave with you; my peace I give unto you!" Yet the same Prince of Peace also declares: "Think not that I am come to send peace on earth: I am not come to send peace, but a sword." The heavenly peace which the Lord sends the good, is the unition of the soul with himself. The world's peace, the infernal peace of the evil, is the obdurate heart, the seared conscience, the quiet mind of the reprobate, against whose repose the Lord sends forth the sword of Divine Truth.

"Honor thy father and thy mother" is a holy precept of the decalogue. What does the Lord mean, then, by that strange declaration, "Whoso hateth not his father and mother cannot be my disciple?" It is not a sufficient explanation to say that this passage only means that we must love father and mother less than we love God. Father and mother, like everything else, may be used in a bad as well as a good sense. The false and evil things in our náture, are descended from or born of the old hereditary masculine and feminine evil principles, which are here denominated father and mother. That evil spiritual father and mother we are to hate.

The Lord always ignores natural things and appearances when they do not represent spiritual verities. He never recognized Joseph as his father or Mary as his mother, because they had no such spiritual relationship to his own spirit. He never called Mary mother but "woman." The Jews were lineally descended from Abraham; but Jesus sternly denies that they are the children of Abraham, and assigns them to their spiritual father: "Ye do the works of your father, the devil."

*Blood* in one sense is the symbol of wickedness, violence, and death. In the other sense it is the symbol of self-sacrifice, infinite love, and the eternal life of divine truth. *Mountains* in one sense mean the supreme or loftiest good affections of the soul, which are said to break forth into singing and gladness at the presence of the Lord. In the opposite sense they are the opposite evil affections, which are to be removed and cast into the sea, or the depths of hell, by means of a living faith.

*Wine* is the precious emblem of divine truth; and the feast which the Divine Wisdom provides for the soul, is described as "a feast of old wines." Yet the awful heresies with which Babylon intoxicated the world, are symbolized also by wine: and the priests of a perishing church are described as erring in vision and stumbling in judgment "through wine and strong drink," which are false doctrines or falsifications of truth.

"Eating and drinking, marrying and giving in marriage," are the terms applied to the wicked and sensual antediluvians, who mocked at Noah's earnest preparations to escape the impending flood. There are two kinds of "eating and drinking." Here it is evidently used in the bad sense. But the solemn declaration, "He that eateth my flesh and drinketh my blood, dwelleth in me and I in him," expresses the most vital truth of religion. One kind of eating and drinking is the appropriation of the evil and false, and their assimilation into the organic life of the soul. In the opposite and good sense, it is the appropriation and assimilation of the good and true.

There is a heavenly marriage and there is an infernal marriage. When a man knows the truth and loves to obey it, he is in the heavenly marriage and wears the wedding-garment. When he believes the false and loves to live in it, he is in the opposite or infernal marriage, and will assuredly be cast "into outer darkness."

When goodness or charity animates the will of man and truth illumines his understanding, and when they act together like man and wife, and good works are their offspring, we have that marriage between the complementary parts of the individual soul which we call regeneration. When there is a similar union of two souls, male and female,—the female element representing the purified affections of the will and the male element representing the truthful thoughts of the understanding,—we have the heavenly marriage between the sexes, which was instituted by the Lord in Eden, which he has dowered with his perpetual blessing upon earth, and which is the central and fundamental form of social life in heaven.

The infernal marriage is the exact opposite of the heavenly. As every good affection has a sympathetic affinity for that thought, or form of wisdom, which gives it the best and happiest expression, so every evil affection has a passional attraction for the specific falsity or error which excuses it, approves of it, loves it, and gives it shape and power. This gradual conjunction or marriage of the evil and the false, mutually attracting and espousing each other, and then prolific of a wicked life, is meant by the "eating and drinking, marrying and giving in marriage" of the antediluvians.

How are we to determine which kind of marriage the Lord meant in the passage before us? Simply as in all other cases—by the context. The character and motives of the persons who put the question, and the principles involved in the case propounded, must determine the true meaning of the response.

Many inquiries were made of Jesus during his earth-life, and his answers were always adapted to the mental and moral states of the questioners. Sometimes, to the natural man's apprehension, he appears to conceal: sometimes, to evade. Sometimes his reply seems to evince only the lower grades of worldly prudence; then gems of spiritual wisdom fall from his lips; and under the highest conditions his answer is a sun-burst of celestial glory. To the greatest of all questions—Whence comest thou? and what is truth? he was profoundly silent: for Pilate, the type of the old, perishing pagan civilizations, could not have received the glorious responses. At the other extreme stands the young man who had kept all the commandments from his youth, and whom Jesus loved; and to him he bequeaths that sentence of transcendent beauty, incomprehensible to the sensual mind:

"One thing thou lackest. Go thy way, sell whatsoever thou hast, and give to the poor, and thou shalt have treasure in heaven: and come, take up the cross and follow me."

The question, Whose wife shall she be? was put by the Sadducees, a party or sect in the Jewish Church which stickled vehemently for a purely literal interpretation of the Scriptures. They denied the resurrection and a future life. It was to throw ridicule on the life after death, that they proposed what they considered an insurmountable difficulty in the event of a woman meeting her seven husbands in another world. They put the question in a hypocritical and mocking spirit. As Dr. Adam Clarke observes, "It was the question of libertines."

Our Lord rebuked their literalism by telling them that they erred greatly, "not knowing the Scriptures, nor the power of God." How could it be otherwise? The power of God is concealed in the spiritual sense of the Scriptures; and it is the naturalism in men, leading them to interpret the Lord's words literally, which has always caused the church to err.

He rebuked their denial of the resurrection by reminding them that Abraham, Isaac, and Jacob had already risen from the dead, and were living in heaven,—a fact of which a great many literalists need to be reminded at the present day.

He rebuked their sensuality by concealing from their gross understandings the most beautiful truths of the heavenly life. There was no ground in such minds ready for their reception. The divine truth is always pressing from heaven to come to us and bless us; but it can come only through human mediums. No medium, no truth; false mediums, false doctrines. To these men nothing could be given but the vague, incomprehensible phrase, "They are as the angels of God."

Examine the case propounded by the Sadducees, and you will discover that the principles involved in its structure stamp it as a picture of the evil or infernal marriage, which cannot exist in heaven, and which indeed effectually and forever excludes from the heavenly life.

It was a marriage in which the prime motive was the propagation of the species; that he might "raise up seed unto his brother." Such things are impossible in the spiritual world.

It was a forced, unnatural union, commanded by law. The parties might have no spiritual affinity or attraction, indeed might hate and loathe each other; and still they were obliged to submit to the usages of the nation. It was purely external.

It was a marriage involving polygamy; for although a man might have a congenial wife of his own, he was compelled to take his deceased brother's wife also to his arms.

It was a marriage so often repeated as to destroy every vestige of true conjugial love in the womanly part of our nature, and to bestialize the affections to the lowest degree.

This marriage, propounded by persons who denied the life after death and the spiritual sense of the Scriptures, was plainly destitute of every element of spirituality.

This marriage, wholly sensual and external, a mere civil or legal alliance with no interior vitality, a marriage which might be indefinitely repeated, which might be coincident with another already existing; this marriage, the proper external type of the union of the evil and the false, and therefore childless and fruitless; this revolting marriage was presented to our Lord by the Sadducees, who did not revolt at it, but regarded it complacently from their sensual stand-point as a fair type of the union of the sexes; and the question was propounded: "In the resurrection whose wife shall she be?"

If the tribute-money which Jesus declared should be rendered unto Cæsar, bore "the image and superscription" of Cæsar, whose "image and superscription" is stamped upon the face of this Sadducean marriage? To whom shall it be assigned? Is the seal of heaven upon it? Is its existence possible among angels?

Our Lord replies:

"The children of this world marry and are given in marriage;

"But they who shall be accounted worthy to

attain that world and the resurrection from the dead, neither marry nor are given in marriage."

And his meaning clearly is this:

Carnal and sensual men, "children of this world," have marriage unions upon a merely external or legal basis, destitute of spiritual life; a kind of union which does not and cannot exist in the spiritual world, where angels who have been raised from spiritual death or have attained unto the resurrection, are united in a different manner, incomprehensible to the present generation of men.

"They are as the angels of God in heaven." He does not tell them who the angels are. He does not unfold the laws of their being or the state of their life, the relation which they hold to each other, nor the outward social and institutional forms which their spiritual affections necessitate. These things were carefully withheld from that sensuous and incapable age. When we understand how it is with the "angels of God," we shall understand the mystery of marriage.

The concealments of Divine Providence are as wonderful and as merciful as its revelations.

The Sadducees represented the whole world at that time on the question of marriage. If the Lord had attempted to explain to them the difference between the marriage of angels and the marriage of the "children of this world;" if He had given them the idea that there was any marriage at all in heaven; they would not have grasped *his* idea, but would have corrupted and falsified it with their own. They would have imagined that angels have sexual passions like ourselves, are attracted and excited by physical grace and beauty, and contract alliances very much as we do on earth; all of which would have been totally false. Such misconceptions might have been of incalculable injury to the infant Church. It was far better to leave them in darkness and error, than to give them a spiritual truth which they could not understand, and which they would have falsified and defiled by sensuous interpretations.

Let us notice how subtly and beautifully the resurrection from the dead is connected with our spiritual idea of marriage. Our Lord always means spiritual things when he seems to speak only of natural things. He means spiritual resurrection. All men undergo the resurrection from the dead—but "the resurrection of life" and "the resurrection of damnation" are opposites. The resurrection of life is the putting away of all earthly and sensual modes of thought and feeling, and the reception of spiritual and heavenly life instead. The poet understood it when he wrote:

> "That men may rise on stepping-stones
> Of their dead selves to higher things."

It is this resurrection by which we cease to be "the children of this world," and become "equal to the angels." It is this resurrection which delivers us from the bondage of the evil or infernal marriage, the union of wicked lusts and false persuasions, so that we rise above it and leave it beneath us for ever, and become capable of the heavenly or angelic marriage.

Observe also that as this spiritual resurrection is quietly going on all the while we are dying to our earthly selves, even during our life in this world, so in the same degree is the heavenly marriage being consummated in the interiors of our spirits. It is thus that resurrection, marriage, regeneration and heaven are spiritually identical. In one sense, therefore, the bare letter is true, that there are no marryings in heaven: for the essentials of marriage, the union of the good and the true in the soul, the germs of regeneration, must be acquired upon earth to be unfolded into beautiful and eternal outward form in the heavens.

Heaven *is* spiritual marriage, and no one can be admitted into heaven until interiorly prepared for that marriage. The outward nuptials in heaven, transcendently beautiful as they are, are only the sign and seal, the ratification, the ceremonial outbirth, of what has already taken place, unconsciously perhaps, in the depths of two blessed and immortal spirits. On earth very frequently the bodies only are married, or the bodies first and the souls afterward. In heaven the souls are married first, on an eternal basis, and the bodies afterward; so that heavenly purity, order, and beauty may descend from the interior life into the outer or bodily life, as the sunlight enters a cloud and colors every atom of its form with its own ethereal and resplendent glory!

The reader may now see that the true meaning of this often-quoted text does not lie upon the surface; that a great deal more is meant than the sensuous understanding apprehends; and that, rationally interpreted, our Lord's words do not teach that there are no marriages in heaven. They really teach this: that men who have attained the angelic state by dying to themselves and the world through obedience to God, have put off or renounced those evil passions and false principles whose union is the infernal or the merely sensual marriage; and have thereby become so different from "the children of this world," so elevated above them, that the same life, fortune, fate, and social forms can no longer be predicated of both parties. The word "marriage," as applicable to one, becomes wholly inapplicable to the other. The angels from their spiritual stand-point, can say with equal truth of "the children of this world," they neither marry nor are given in marriage.

The Lord's words also leave the entire question of angelic forms, affections, and unions untouched and their phenomena unrevealed. If it be his will to unfold the mysteries of angelic life, including those of marriage, to a more receptive age, that revelation must be judged of by its own intrinsic character; and the merely literal sense of this passage, which has truly killed its spirit or real meaning, cannot be brought against it as a legitimate objection.

Why, then, says the reader, have so many wise and good men and women been permitted for so many centuries to entertain such erroneous ideas about heaven and angels, and about the true nature and final social destiny of their own souls? Because truth cannot be clearly revealed to imperfect men and to an infant Church. There is indeed a grand, orderly, progressive evolution and revelation of divine truth. The Church like the race and the individual has its infancy and its childhood; when it is fed upon the milk and not the meat of the Word; when it apprehends all things sensuously and not intelligently; when it is taught by dictations and not by such philosophical explanations as are afterward required. The Divine Word also, like the created and unwritten word of nature, is a vast system of symbols or hieroglyphics; some of which are perfectly plain or easily deciphered, especially those essential to moral and spiritual life; while others are very obscure and difficult, demanding for their solution a high degree of scientific and rational culture.

Man at first believes what his senses teach him; and he learns very slowly to distrust the evidence of those most fallacious guides. For how many ages did men believe that the sun moved round the earth, rising and setting as it seems to do! And blinded by theological prejudice, how absurdly did they resist the scientific correction of the false report of their senses! Nature has already revealed to our modern analysis, things which would have been more wonderful to our ancestors than all their fables and myths and fairy-tales seem to us. The sensuous interpretation of nature ceases with the advent of true science and philosophy. And since the Word of God and his creation are analogous, producing similar effects upon the human mind, we see that the Church has had its period of literal or sensuous interpretation, to which it still clings with a childish tenacity, ignoring the happy dawn of a truly rational spirit, which will compel it to put away childish things.

Marriage in heaven is denied solely in the spirit and on the strength of the letter. Another and instructive example of false doctrine arising from sensuous or literal interpretation, is the confident belief of the apostles and early Christians that the second advent of the Lord was close at hand. Jesus had declared that that generation should not pass away, before his description of the last judgment would be fulfilled. He had also permitted them to infer that the Apostle John should tarry until his second coming. And the angel commands the Seer of Patmos not to seal up the prophecy of the book, with the ominous declaration, "for the time is at hand." The spiritual and real signification of all these things was not understood. False doctrines based on the apparent truth of the letter spread through all the churches; and every century since the time of Christ, has had its grand agitation on the Second Advent, and the same cruel disappointment, to the delight of scoffing infidels.

If the articles of the present Christian faith are critically examined, it will be seen that this principle of sensuous or literal interpretation has tainted the whole body of religious doctrine, so that not one stone of the temple of truth is left upon another in the relations of spiritual order and beauty they were designed by the great Architect to occupy. The history of the Church is a history of the struggles between the unyielding spirit of Literalism and the growing and expanding spirit of Rationalism, discovering the untenableness of the letter, and attempting to spiritualize it by its own unaided imagination. It is evident to thinking men that the Church must outgrow the limitations of the letter, or the human mind will outgrow the Church. It must spiritualize with Swedenborg, or it will inevitably rationalize with Strauss and Rénan.

The first stage of the life of the Church is like that of the child; and the first duty, to learn the Father's will and do it without questioning his authority or prying into the uses and philosophy of his system of government. He is taught by dictation on the plane of the intellect, and led by love on the higher plane of the affections. If he attempts to reason, his powers are inadequate, his data insufficient, and his conclusions necessarily fallacious. In the after-stages of his development, he is not only permitted but invited to cultivate and use his own powers of observation and analysis. With expanded faculties, new data, new liberty of thought and action, he penetrates mysteries before inscrutable. He then corrects, in theology as well as in science, the erroneous impressions of his senses, and by the opening of his spiritual perceptions he participates with the angels in the knowledge of genuine Truth.

"I have many things to say unto you, but ye cannot bear them now," said Jesus to his disciples; and although he has made glorious additions to the store of truths then given, the words are still true and will forever remain so. The truths revealed by each New Dispensation, however great, are few and small in comparison with those which must forever remain concealed in the bosom of God. One age or Church is ready to hear what a preceding one could not receive. Spiritually speaking, Swedenborg's doctrines are nothing new, but only the genuine growth and outflowering of the old, eternal truths of the spiritual universe. Through him the Lord has given us many, many things which the Apostolic Church could not bear. One of the most beautiful and wonderful of these divine gifts, is a true doctrine of sex, love, and marriage, the great keystone of the spiritual arch. Under the narrowing influence of literalism, the Christian world may for a while regard it with incredulity or aversion, as it did the astronomy of Galileo and the discoveries of geology; but the spirit of truth, which conquers and reconciles and unites all things, will finally overcome the prejudices and dissipate the darkness of the human mind.

All things, then, are bi-sexual; God; Man; Nature; the Word; the Church. Nothing exists but by the reciprocal attraction and inspiration of two coequal forces or powers. Male and female or sex, attraction or love, equilibrium or marriage, is the key to all science, spiritual and natural. Our forms, our loves, our unions, are imperishable. Nature affirms it; the spirit reveals it; our Lord does not deny it.

That tender, holy, heavenly conjunction of two hearts into one, which is effigied in the marriage of the will and understanding of the regenerate soul, and which is itself a miniature of that infinite marriage between the Lord and his Church, is the eternal source of our life and joy. For death is a shadow; we never cease to live or cease to love.

"Our echoes roll from soul to soul,
And grow forever and forever."

## CHAPTER IV.

### *WHAT SWEDENBORG SAYS ABOUT IT.*

THEY are as the angels of God in heaven."

Sublime, mysterious words!—the embodiment of all we can imagine of innocence, beauty, peace and love; suggesting a realm of unimaginable glory and wonder; farther off, it seems to us, than Orion or the Pleiades; more inaccessible than the golden islands and seas of a summer sunset!

And yet—

Are we *never* to know *how* it is with the angels? Does the impassable gulf which separates heaven from hell, stretch its awful abysses also between heaven and earth? Then how did Moses and Ezekiel and John and Paul see into heaven? And what does our Lord mean by those strange and hope-inspiring words?—

"Hereafter ye shall see heaven open."

More than a century ago a Swedish philosopher of great learning, genius and virtue, professed to have the sight of his spirit opened, whereby he was intromitted into the spiritual world. He remained in this state for more than a quarter of a century, during which he studied the laws and phenomena of that world, and unfolded a complete system of psychology and theology, which he declared to be revelations from heaven for the use of a New Church signified by the New Jerusalem in the Apocalypse. A strange man he was, and the author of strange books; a psychological problem as yet unsolved. He made no proselytes, founded no Church, sought no distinction; but dropped his seed into the field of theological literature, as quietly and patiently and with a faith as unwavering as a man who plants an acorn and sees with the mind's eye his grandchildren playing in the shadow of the mighty oak.

"A revelation from heaven!" exclaims the incredulous child of modern philosophy—as if all things were not revelations from heaven!

Heaven? The flowers reflect it; the birds sing of it; the ocean voices it; the stars point to it; the winds breathe it; the sun blazons it, all nature mirrors it! Our human souls are full of it; our thoughts flow from it; our loves are born of it; our music, architecture, poetry, art, science and all, are the falling shadows of its transcendent glory!

Revelations or communications from the spiritual to the natural sphere, have been given in all ages and to all peoples and tongues, varying in their form, extent, and character according to the nature, circumstances, and receptivity of those to whom they were sent.

The highest revelation was "the voice of the Lord God walking in the garden." Then came special visitations and oral instruction by angels. Then the inspiration of a written Word containing Divine Wisdom in its unseen but infinite depths. Sometimes revelations have been made by miracles, sometimes by dreams, and sometimes by open vision into heaven or the world of spirits.

A new Church and a new revelation come, not because the preceding were false or had failed of their mission, but because new, higher, larger, better conceptions of truth have become possible and necessary by the gradual and orderly unfolding of the human mind. Our spiritual life always depends upon our communication with heaven; and the form of the revelation depends upon and varies with our mental states. And all revelations, however diversified in external form, have the same interior source and life. There is no antagonism between what Jehovah taught in the garden and what Moses received on the mount. The ray of divine light is continuous through Psalmist and Evangelist, through Prophet and Apostle. And Swedenborg in his highest flight, only interprets the old song which Moses and John, "like poets hidden in the light of thought," have always been singing, unheard by the world.

Revelations are authenticated, not by miracles, nor by the authority of great names, nor by external evidence of any kind, but solely by the luminous and beneficent truths they convey to man. Revelation reaches the first or lowest stage of human thought and life by miraculous impression upon the senses: it reaches a higher and nobler stage by rational illumination of the understanding. Yet no revelation is ever recognized as such by those who look from the old standpoints, and who have not interiorly outlived and outgrown the peculiar spirit and forms of the preceding dispensation.

A rational age does not ask for miracles, but for light. Swedenborg's system must stand, not upon his mere *dicta*, but on its own intrinsic merits. Is it rational, satisfying the understanding? Is it beautiful, satisfying the æsthetic longings of the soul? Is it consistent with the Sacred Scriptures, satisfying our ineradicable loyalty to Revealed Truth? It is only by this method that we can assure ourselves whether this new doctrine is the genuine light of heaven, or some mysterious flash of magnetic fire from the northern skies.

Whatever may be the verdict of posterity, there is unquestionably an increasing and legitimate curiosity among intelligent people to know what Swedenborg said about the spiritual world. We will gratify our readers and illustrate our subject by some quotations from this writer, who, entirely aside from his theological pretensions, has made some of the profoundest observations on the nature of the human soul, the origin of love and the philosophy of marriage. If he did not get his singularly beautiful doctrines from a spiritual source, he is at least the founder of a system of psychology which is more in harmony with the rich discoveries of modern times in all departments of science, than any of the systems which have engaged the attention or secured the faith of men.

Communication with the spiritual world is commonly regarded as impossible or improbable, because men are in the densest ignorance of the true nature of spirit, and of the connection between heaven and earth. They cannot imagine how a man living on earth can see into heaven; and yet if *that* be impossible, the Bible is false and revelation a dream. Swedenborg explains how this seemingly miraculous insight into the spiritual world is effected. He says:

"Angels cannot be seen by man with the eyes of his body, but only with the eyes of his spirit (or spiritual body) which is within him.

This spiritual body communicates with the spiritual world, while all parts of the natural body are in the natural world. Like sees like, because from a like ground. Objects which are above the sphere of nature, as all those of the spiritual world are, may however be seen by man when he is withdrawn from the sight of his body, and that of his spirit is opened. This is done in an instant when it is the pleasure of the Lord that the things of the spiritual world shall be seen by man: nor is he at all aware at the time that he does not behold them with his bodily eyes."

"It was in this way that angels were seen by Abraham, Lot, Manoah and the prophets. It was in this way that the Lord was seen by the disciples after his resurrection; and it was in the same way also that angels have been seen by me.

"As the prophets enjoyed this mode of vision, they were called Seers, or *the men whose eyes were open;* and to cause them to see in this way was called *opening their eyes.* This was done to Elisha's servant, of whom we read:

"'And Elisha prayed and said: Jehovah! I pray thee, open his eyes that he may see.

"'And Jehovah opened the eyes of the young man, and he saw; and behold! the mountain was full of horses and chariots of fire round about Elisha.'"

The process by which Swedenborg saw spirits and angels being thus rationally explained, we may credit his assurance that all spirits are in the perfect human form. He says, moreover, that they are male and female; that they are animated by loves, feelings, and sentiments similar to those which people on earth experience. Societies are arranged in that world, not geographically, nor by any external law, but by the great internal law of spiritual affinity. Similar or sympathetic men and women are drawn together. The male and female souls most congenial, most sympathetic, are united in an eternal union, which is the marriage of heaven.

His picture of the heavenly societies is surpassingly beautiful. The perfect order, peace, beauty, love, which there reign supreme, would realize our grandest and holiest ideal of the final home and rest of the saints. We are at first surprised and bewildered, but finally convinced and delighted, as he unfolds to us the wonderful fact, that the secret principle and key to all this heavenly organization is the conjugial love. Sex belongs to the soul as well as to the body: and, as our souls are immortal, sex, with all its aspirations, is eternal. Regenerate souls united forever in conjugial love, constitute the Lord's church or kingdom in heaven.

Of this conjugial love, the connecting bond of angelic hearts and the life of the heavenly home, Swedenborg makes the following general statements, which may serve to show how important a place it occupies in the psychology and theology of the New Church:

"There is a love truly conjugial, which at this day is so rare that its quality and almost its existence are unknown.

"It originates in the marriage of goodness and truth from the Lord in the regenerate souls of man and woman, and corresponds to the spiritual marriage of the Lord with his church.

"From its origin and correspondence, this

conjugial love is celestial, spiritual, holy, pure, and clean above every other love imparted by the Lord to the angels in heaven and to the men of the church.

"It is the fundamental love of all celestial and spiritual loves, and thence of all natural loves; and into it are gathered all the delights and joys of the human soul.

"It belongs to the internal or spiritual man; and none come into this love, or can remain in it, but those who love the Lord and obey his commandments."

How the natural love of the sex, common to man and animals, is changed into this pure, holy, and perfect love, is also explained in his writings; from which we may learn that regeneration, resurrection, marriage, and heaven, are in their spiritual sense synonymous terms. We are thus led into an entirely new sphere of psychological and theological truth.

We are aware with what incredulity and distrust these revelations will be received. Man is slow to discover the truth, and even slower to recognize it after it has been discovered. We plead for free thought, charity, and liberal construction. Aside from its psychological and theological issues, this doctrine of love and marriage in heaven has an æsthetic value, sure at last to command the admiration of men. We will show the practical tendency of these novel teachings, and satisfy a legitimate curiosity, by drawing from Swedenborg some descriptive illustrations of the highest state of marriage and conjugial love in heaven; for he has a charming method of making his philosophical principles objective and intelligible by narrating scenes and conversations with spirits and angels, which impress them on the imagination and memory. The scene of the following narration is in the world of spirits, a realm between heaven and earth, or the region or state which every one enters immediately after death:

"On a time when I was meditating on conjugial love, lo! there appeared at a distance two naked infants with baskets in their hands and turtle-doves flying around them. On a nearer view, they seemed as if they were naked, but beautifully ornamented with garlands. Chaplets of flowers decorated their heads, and wreaths of lilies and roses, hanging obliquely from the shoulders to the loins, adorned their bosoms. Round about both of them there was, as it were, a common band woven of small leaves interspersed with olives. But when they came nearer they did not appear as infants or naked, but as two persons in the prime of life, wearing cloaks and tunics of shining silk embroidered with the most beautiful flowers. When they came near me, there breathed forth from heaven through them a vernal warmth attended with an odoriferous fragrance, such as arises from gardens and fields in the time of spring.

"They were two married partners from heaven, and they saluted me; and because I was musing upon what I had seen, they inquired, 'What did you see?' When I told them that at first they appeared to me as naked infants, afterward as infants adorned with garlands, and lastly as grown-up persons in embroidered garments, they smiled pleasantly and said:

"In the way we did not seem to ourselves as infants, or naked, or adorned with garlands, but constantly in the same appearance which we now have."

They then explain to Swedenborg how the beautiful picture first presented, made up of naked infants and doves and uniting wreaths of flowers, was emblematical of the innocence, sweetness, peace, purity, and exquisite delights of their conjugial love. In the spiritual world every one's sphere precedes him, and tells in charming symbolic statuary or painting the character and peculiarities of the one who approaches him. There as here, "distance lends enchantment to the view." In the same manner also every beautiful and useful object upon earth, however ignorant we may be of it, represents the sphere or outflowing of some spiritual truth or some heavenly affection.

They also explain to him the cause of the vernal glow and the exquisite fragrance of fields and gardens which surround them. They tell him it comes from the constant spiritual spring in which they live; for love and wisdom, which are spiritual heat and light, are united in their minds in equal proportions. They assure him also that it is the same conjugial sphere descending from heaven to earth, which causes the germination of leaf and blossom and the connubial associations of birds and animals, as well as the tender gushes of sentimental passion in the hearts of human beings.

> "In the spring a fuller crimson comes upon the robin's breast;
> In the spring the wanton lapwing gets himself another crest;
> In the spring a livelier iris changes on the burnished dove;
> In the spring a young man's fancy lightly turns to thoughts of love."

"He gave me his right hand," continues Swedenborg, "and conducted me to houses inhabited by married partners in a like prime of life with himself and his partner; and he said:

"These wives who now seem like young virgins, were in your world infirm old women; and their husbands who now seem in the spring of youth, were in the world decrepid old men. They have all been restored by the Lord to the prime of life, because they mutually loved each other, and from religious principles shunned adulteries as enormous sins."

He gives the following beautiful description of an angelic pair from the third or celestial heaven,—that heaven which Paul visited, but whose wonders he was not permitted to reveal:

"'One morning I was looking upward into heaven (from the world of spirits) and saw over me three expanses one above another. I saw that the first expanse which was nearest, opened; and presently the second which was above it; and lastly the third which was highest. I then perceived that above the first expanse were the angels of the first heaven; above the second expanse were the angels who compose the second heaven; and above the third expanse were the angels who compose the third or highest heaven.

"'I wondered at first what all this meant; and presently I heard from heaven a voice as of a trumpet, saying,

"''We perceive that you have been meditating on Conjugial Love: and we are aware that no one on earth knows what conjugial love in its origin and essence is. Yet it is of importance that it should be known. It has therefore pleased the Lord to open the heavens to you, in order that illustrating light may flow into the interiors of your mind. Our delights in the third heaven are principally derived from conjugial love. In consequence, therefore, of leave granted us, we will send down to you a conjugial pair for your inspection and observation.'

"There instantly appeared a chariot descending from the highest or third heaven, in which I saw at first one angel; but as it came nearer I perceived that it contained two. (Angelic married pairs always appear at a distance as one person or angel.) The chariot afar off glittered before my eyes like a diamond, and to it were harnessed two young horses as white as snow. Those who sat in the chariot held in their hands two turtle-doves, and they called out to me, saying:

"'Do you wish us to come nearer to you? If so, take heed lest the flaming radiance of the heaven from which we have descended penetrate too interiorly. It would illuminate, indeed, the ideas of your superior or interior mind; but those ideas are ineffable and incommunicable in the world in which you dwell. Therefore let what you are about to receive enter only your rational or lower mind, so that you can make your fellow-men understand it."

"I replied, I will observe your caution: come nearer."

There is a great psychological truth hidden in the above paragraph. The human mind itself has three degrees, expanses, or heavens, one above another, the opening of any one of which intromits us into the heaven of that degree. Swedenborg's own mind was opened to the celestial degree on this occasion, and the angels appeared to him to descend only because he ascended. The things in these degrees are so discrete that what is seen or heard in one may not be understood or remembered when the perceptions of the soul ascend or descend into another degree. Paul could not remember or explain what he heard when in the celestial degree, because his natural degree was closed at the time. When he returned consciously into that degree, the celestial was closed and all its glories had vanished. Swedenborg is warned to receive and retain in his lower mind the angelic truths, so that he could communicate them to us who live on that plane. He concentrates his intellect upon the rational meaning of what he is about to hear; and continues his story:

"So they came nearer; and lo! it was a husband and wife. They said: 'We are a conjugial pair; we have lived happy in heaven from the earliest period, which you call the golden age; and have continued all that time in the same bloom of youth in which you see us now.'

"I viewed each of them attentively, because I perceived that they represented conjugial love in its life and in its decorations; in its life by their faces, and in its decorations by their raiment. All angels are affections of love in the human form. The ruling affection itself shines forth from their faces; and from the affection the kind and quality of their raiment is determined. Therefore it is said in heaven that every one is clothed by his own affection."

Swedenborg then minutely describes their appearance and dress, every point in which was, according to the creative laws of the spiritual world, symbolical of their own spiritual qualities. He labors in vain to do justice to beings so transcendently beautiful. We gather, however, from his description, that the face of the husband was "one resplendent comeliness," and that the beauty of the wife was immeasurably beyond the reach of pen or pencil. He gives us a touch or two worthy of Dante for poetic charm and grandeur.

"I saw her face, and I did not see it. I saw it as essential beauty; and I did not see it, because this beauty was indescribable. In her face was a splendor of flaming light, such as the angels in the third heaven enjoy; and this light made my sight dim, so that I was lost in astonishment. Observing this, she addressed me, saying, 'What do you see?'

"I replied: 'I see nothing but conjugial love and the form thereof: but I see and I do not see.'

"*Hereupon she turned herself sideways from her husband, and then I was enabled to view her more attentively.*'

Dante acquired strength to look at the sun by gazing on the heavenly face of Beatrice. The Dantes and Beatrices of the angelic spheres, acquire strength to look upon the Lord by gazing into each other's faces: but then their glory becomes blinding to earthly visitants. They can diminish the flood of celestial light which surrounds them by turning away from each other, and thus unveil their beauty to mortal eyes.

Swedenborg continues in a strain almost as beautiful:

"Her hair was arranged with a taste corresponding to her beauty, and in it was inserted a diadem of flowers. She had a necklace of carbuncles from which hung a rosary of chrysolites, and she wore pearl bracelets. Her robe was scarlet, and fastened in front with a clasp of rubies. What surprised me in all this was, that *the colors varied according to her aspect in regard to her husband, being sometimes more glittering, sometimes less.* If she looked toward him, they shone more brilliantly: if she turned sideways, they shone less."

A similar illumination of precious stones by influx of light or truth from the Lord, was the secret of the Urim and Thummim, or twelve precious stones worn on the breastplate of the Jewish high-priest. They gave responses by brilliant variegations of color, a veritable language of light, as intelligible as the language of flowers. Josephus affirms that their light was perfectly dazzling when Jehovah promised victory to the Jewish armies. This wonderful property of the breastplate has always been a great puzzle to the commentators. Infidels and naturalists deny the facts: Swedenborg alone explains them. The miracles of the Bible are illustrated and confirmed by the simplest laws and phenomena of the spiritual life.

While these heavenly beings are conversing with Swedenborg on the mysteries of conjugial love, he makes a curious observation about their unity or simultaneousness of thought and speech,—a phenomenon impossible in our natural state of life, and one of the results of the perfect marriage union of souls.

"When the husband was speaking, he spoke at the same time as if from his wife; and when the wife was speaking, she spoke at the same time as if from her husband. This union of expression came from their perfect mental harmony. On this occasion, also, I detected the tone of voice which indicated conjugial love.

It was interiorly simultaneous, and it proceeded from the delight of an interior state of innocence and peace."

Of this wonderful and perfect union of two hearts and minds, which is the key to heavenly felicity and all its most beautiful outward accomplishments, our author thus speaks:

"In heavenly marriage there is no domination exercised by one party over the other; for the will of the wife is also that of the husband, and the understanding of the husband is also that of the wife. One loves to think and to will as the other does; and they do so mutually and reciprocally. The result is their final conjunction into one spiritual life. This conjunction is actual; for the will of the wife enters into the understanding of the husband, and the understanding of the husband into the will of the wife, more especially when they look each other in the face; for as has been often before stated, there is an actual transference of thoughts and affections in heaven."

How have these glorious angels, once men and women like ourselves, attained to this exalted state, so that they think, feel, and act sim ultaneously? that they appear to others at a distance as one person? that when one speaks, it is seen and felt that both speak? that they have forms of such transcendent grace and beauty, radiant with celestial love and wisdom? and that all objects around them are warmed and colored with the sweet spiritual life that wells up from their own happy bosoms?

This is the answer:

They are "the children of God, being the children of the resurrection." They are dead to every selfish feeling and thought; dead to every carnal desire and motive; dead, utterly dead to everything false and evil. They have risen to spiritual life in God; to eternal holiness and purity; to the spiritual marriage, incomprehensible to the carnal man, and for ever concealed from his eyes; and in the seraphic calm of mutual love and wedded bliss, they are images and likenesses of the Heavenly Father.

Would not these "children of the resurrection," with their super-sensual and holy ideas of marriage, forgetting all the evil and pain of their earth-life, looking by chance downward and backward into our dark and unholy and selfish souls, poor "children of the world" as we are—would they not say pitifully of us, as we say blindly of them :—

"They neither marry nor are given in marriage?"

Trusting that the reader will see in these descriptions, and in the principles involved, something more than the play of fancy or imagination ;—will see, indeed, that with supreme art the great philosophic artist is simply clothing the True in the garment of the Beautiful ; we will add for his instruction the following account of a marriage-ceremony in heaven given by this illustrious seer.

A marriage-ceremony in heaven!

Yes :—and is it not as rational and probable as the tabernacle seen in heaven by Moses, or as the still more remarkable objects and events there seen and described by the prophets? If angels are male and female ; if all the sweet and gentle and heroic sentiments and passions survive death and exist forever ; if they live in houses and homes, and under perfect civil and social institutions ; if marriage is perpetuated in the holiest form as the eternal type of the Lord's mystical union with his Church; is it not most probable and proper that, when two conjugial partners, having passed through all the exploratory processes of the judgment, are brought face to face, each being perfectly revealed as the other's ideal and counterpart, there should be some public and official recognition of the great fact, some high and holy and final seal set to a union provided and blest of God, and which is to be thenceforth eternal?

Surely no event or usage in either world can have a grander significance than this—the culmination of spiritual affinities, the crowning work of love, in which the Heavenly Father confirms and blesses the perfect and eternal union of two of his redeemed and happy children! Indeed, the subject is so transcendent and holy, that Swedenborg eschews the futile graces of poetry and rhetoric, and treats it with the austere simplicity of antique art.

"Toward evening there came a messenger clothed in linen to the ten strangers who attended the angel [newly-arrived spirits from the earth], and invited them to a wedding which was to be celebrated the next day; and the strangers were much pleased to think that they were to witness a marriage ceremony in heaven.

"When they awoke in the morning, they heard the singing of the virgins and young girls from the houses around the public places of resort mentioned above. They sang that morning the affection of conjugial love, the sweetness of which so affected and moved the hearers, that they perceived a blessed serenity instilled into their bosoms.

"At the hour appointed the angel said: 'Make yourselves ready, and put on the heavenly garments which our prince has sent you.' They did so; and lo! the garments immediately became resplendent as with a flaming light. They asked the angel, 'What is the reason of this?' He replied, 'Because you are going to a wedding; and when that is the case, our garments always assume a shining appearance, and become wedding-garments.'

"After this, the angel conducted them to the house where the nuptials were to be celebrated, and the porter opened the door. They were received and welcomed by an angel sent from the bridegroom, and were introduced into an antechamber and shown to the seats intended for them. There was a table in the middle of the room, and on it a magnificent candlestick with seven branches and sconces of gold. Against the walls were hung silver lamps, which being lighted made the atmosphere of the room of a golden hue. There were two other tables on which loaves of bread were placed in three rows; and each corner had a little table with crystal cups upon it for wine.

"While they were looking at these things, a door opened near the marriage-chamber, and six virgins came forth; and then the bridegroom and the bride, holding each other by the hand. The angelic pair advanced and took seats opposite the candlestick, the bride on the right and the virgins standing along at her side. The bridegroom wore a mitre on his head, and was dressed in a bright purple robe and a tunic of shining linen, with an ephod on which was a golden plate set round with diamonds. On the plate was engraved a young eagle, the marriage ensign of that heavenly society. The bride was dressed in a scarlet robe, underneath which was a garment ornamented with fine needlework. Beneath her bosom was a golden girdle, and on her head she wore a golden crown set with rubies.

"When they were seated, the bridegroom, turning himself to the bride, put a golden ring on her finger. He then took bracelets and a pearl necklace, and clasped the bracelets about her wrists and put the necklace around her neck, and said: '*Accept these pledges.*' When she had accepted them, he kissed her and said: '*Now thou art mine;*' and he called her his wife. Then the company exclaimed: 'May the divine blessing be upon you.' These words were first pronounced by each one separately, and then by all together. They were pronounced also in his turn by a certain person sent from the Prince to be his representative. At that instant also the antechamber was filled with an aromatic smoke, which was a token of blessing from heaven.

"Then the servants in waiting took loaves from the table near the candlestick, and cups now filled with wine from the tables in the corners of the room, and gave to each of the guests bread and wine; and they ate and drank. After this the husband and wife arose, and the six virgins, holding silver lamps in their hands, attended them to the threshold, and the married pair entered their chamber."

Every object and circumstance of this simple ceremony, was emblematic of something spiritual. The golden candlestick, the silver lamps, the shining garments, the auroral atmosphere, the six virgins, the dress of the bride and bridegroom, the pledges given and accepted, the blessings bestowed,—the whole scene in its greatest and least parts, had sacred meanings to the illumined assembly, which in our highest states we can but partially understand.

The grandest and most solemn feature in this ceremony was the partaking of bread and wine as in the Holy Supper. Marriage, which typifies regeneration, and the Lord's church or heaven in the soul, is also typified by the Holy Supper. The final marriage of two regenerate souls in heaven, is the highest and holiest act of union with the Lord, the genuine spiritual supper, whereby He enters into and sups with them and they with Him.

"'Then a certain wise personage, one of the marriage-guests, said to the strangers: 'Do you understand the meaning of what you have seen?' They replied: 'Only a little of it.' And then they asked him why the bridegroom was dressed in that singular manner. He answered: 'Because the bridegroom represented the Lord (who is our great High Priest), and the bride represented the Church; for *marriages in heaven represent the marriage of the Lord with the Church.* This is the reason why he wore a mitre on his brow, and was dressed in a robe, tunic, and ephod like Aaron; and why the bride had a crown on her head and wore a scarlet mantle like a queen. To-morrow, however, they will both be dressed differently, for this representation lasts only a day.

"The strangers asked: 'Why were there not bridesmen with the bridegroom to correspond with the virgins who attended the bride?'

"The wise one answered: 'Because to-day we males are numbered among the virgins.'

"They said: 'Explain your meaning.'

"He replied: 'A virgin signifies the Church, and the Church consists of both sexes. Therefore also we men, *with respect to the Church*, are virgins. This is evident from these words in Revelation:

"'These are they who were not defiled with women: for they are virgins; and they follow the Lamb whithersoever he goeth.'"

The poets, those child-like interpreters of spiritual truth, understand perfectly that personal purity and chastity are not lessened, but immeasurably augmented, by marriage. Coventry Patmore, the poet of conjugial love, interprets the Revelation better than all the professional commentators:

> "Virgins are they before the Lord,
> Whose souls are pure. The vestal fire
> Is not, as some mis-read the Word,
> By marriage quenched, but burns the higher."

"Lastly, they asked: 'Is it not expedient that a priest should be present and minister at your marriage-ceremonies?' The marriage-guest answered: 'This is expedient and proper on earth, but not in heaven, on account of the representation of the Lord himself and the Church. On earth they know nothing of these things. Yet even with us a priest ministers in whatever relates to betrothings, and hears, receives, confirms, and consecrates the consent of the parties.'"

What a vast change it would make in the social condition of our world, if this custom of the spiritual life existed here also; if marriage was regarded as such a holy and divine institution, that none would enter it until they had confided their loves, their hopes, their motives, their aspirations, to a pastor's spiritual and loving inspection; and until every unworthy thought and feeling had, through his assistance, been scourged from their souls!

It is necessary to impress it upon our natural minds, steeped in sensualism as they are, that these heavenly marriages are so chaste, that the life of such marriages is chastity itself. The faintest breath of earthly passion would stain forever the angelic mirror of conjugial purity. By the great law, that similar natures attract, while dissimilar ones repel, a separation is effected between souls after death. The evil pass into evil societies; the good ascend into heaven. The good become continually better and purer and holier. The marriage sphere of heaven is so chaste and powerful, that evil and licentious spirits are blinded and suffocated when they approach the Happy World. Female angels are intensely sensitive thermometers to the heat which arises from the activity of sensual passion. The most hypocritical spirits cannot long conceal their interior licentiousness from such amazingly acute perceptions.

Robertson, a charming English divine, thus illustrates this remarkable detective power which innocence and purity display in the presence of their moral opposites:

"Purity can detect the presence of the evil it does not understand. As the dove which has never seen a hawk trembles at its presence; as the horse rears uneasily when a wild beast unknown and new to it, is near; so innocence understands, and yet understands not, the meaning of the unholy look, the guilty tone, the sinful manner. It shudders and shrinks from it by a power given to it, like that which God has conferred on the unreasoning mimosa."

Most persons to whom the idea of sex and

marriage after death is new and strange, immediately ask the same question as did the Sadducees, What will they do who have been married several times? There is no difficulty on this point, when we reflect that the unions formed in this world from various external and selfish motives, are not marriages at all in the spiritual sense of the word. Most marriages here are alliances or contracts for this world only. They are, however, schools and means of discipline for the spiritual life, making a spiritual marriage possible. God's marriages are spiritual, and therefore indissoluble. Those whom He hath joined together in eternal union by an organic and mutual affinity, can never be sundered—no, not even by death.

All external bonds are broken by death; all human institutions, civil and social, vanish away. In the spiritual world our associations will be determined, not by outward things as on earth, but by our interior spiritual affinities. Those who are similar to us will undergo experiences similar to our own. It is by this great law of passional attraction that societies in heaven and hell are formed, consisting of spirits, male and female, who harmonize so thoroughly in affection and thought, that they must needs live together and have the same objective scenery around them. To every male spirit in heaven, there will be some female spirit more interiorly drawn than any other spirit in the universe, his exact counterpart, his spiritual complement. They will live in the same society and in the same house; have the same loves and the same life; and their union will be necessarily eternal.

No one can know while on earth, whether an existing legal marriage is a true spiritual marriage or not. To judge by appearances is not righteous judgment. Our interior character is sometimes vastly different from our exterior. We cannot truly know ourselves or each other until we are stript of our earthly wrappages, and stand forth in our spiritual character; until what has been spoken in the closet shall be proclaimed on the house-tops. Swedenborg says that partners who seem exteriorly to be the closest friends, are sometimes interiorly the most inveterate enemies. On the other hand, couples who seem to be very dissimilar and unhappy here, may find, after death and the judgment, that interior and hidden sympathies stream forth and melt them into blissful and eternal reconciliations. So marriage on earth is not a finality. It was meant to be a grand discipline of life, a stepping-stone, a preparation for true marriage hereafter.

It is clear from this, that marriages in this world interpose no difficulties in the way of spiritual marriage in the next. It is asked also, What will the unmarried do? They are drawn to their spiritual affinities, as all are; and sometimes, perhaps, more easily than those who have been externally married and are compelled to divest themselves of earthly affections and memories. Swedenborg says that the Lord in his mercy always provides "similitudes," or conjugial partners, for those who desire love truly conjugial: if not on earth, owing to numberless hindering contingencies, yet He surely provides them in heaven.

Those who have renounced the world and its temptations, and vowed perpetual virginity; neglecting natural duties, and devoted wholly to spiritual things; expecting, from their superior purity and sanctity, a special reward hereafter; are most to be pitied of all the spirits who come into the heavenly kingdom. They are kindly received by all the heavenly societies; but when they feel the sphere of conjugial love, which is one of intense activity, buoyancy, and joy, they secretly revolt at it on account of their false ideas of chastity and holiness, and become sad and fretful. They then go from one society to another, disappointed and doubting, and are only satisfied when they collect together with their like, and enjoy a communion of thought which is repulsive to all truly regenerate spirits. The good among them are finally delivered from their delusions, and find their conjugial mates.

Having convinced ourselves by a rational process that the sexual relation is universal and eternal; that love in this world and the next is the central, unitizing, organizing element in social life; that marriage is the highest, holiest, and perpetual type of those divine forces which create and bind all things together; we are prepared to make a practical application of the great doctrines of Swedenborg on these subjects.

The most momentous questions before the world, are not those of government, or finance, or constitutions, or law, or science, or art, or physics, or metaphysics; but these:

What are the spiritual differences between man and woman?—or the Philosophy of Sex.

What are the laws of the action and re-action of the sexes on each other?—or the Philosophy of Love.

What is the divine order or plan of union between the sexes, whereby two conjugial partners are made one, creation perpetuated, regeneration effected, and heaven filled with angels, once men?—or the Philosophy of Marriage.

What are the distinctive spheres of life, function, duty, and use for the masculine and feminine faculties?—or the elements of a new, perfect, and final Sociology for man.

We appeal in vain, for answers, to the sciences which are based on the experience of our fallible senses; which practically divorce God from nature, and spirit from matter; and which offer no philosophical bond of connection between the visible and the invisible worlds.

We appeal in vain to a merely literal and

sensuous interpretation of the Word of God, separate from its living spirit;—a system of interpretation which falsifies the Scripture, darkens the understanding, and has been the prolific source of error and mysticism.

We appeal in vain to our short and imperfect human experience, and to the chaos we call history, for any genuine truth about problems which demand for their satisfactory solution the revealing light of heaven.

Swedenborg alone of all men claims to have lived for nearly thirty years in sensible contact with *both* worlds, analyzing both and discovering the relations they hold to each other. Let us see what *he* says about the spiritual differences between man and woman, the philosophy of beauty and love, and the spiritual and eternal uses of marriage. He may give us such rational insight into the organization of the human mind, that we may consider possible, or even probable, his exquisite ideal of conjugial love in heaven, which is so transcendently pure and beautiful, that with our present light we are tempted to regard it rather as the dream of a Poet than the vision of a Seer.

## CHAPTER V.

### *SPIRITUAL DIFFERENCES BETWEEN MAN AND WOMAN.*

THE difference between the spiritual constitutions of man and woman, is fundamental and organic. It is the key to the forms, functions, powers, duties, and peculiarities of each sex; the key to the influence they exert on each other; the key to the spiritual philosophy of marriage, and to many of the most wonderful secrets of nature and man.

Swedenborg has given us this key.

This discovery was not made by scientific induction, nor was it merely the result of rare philosophic insight. It was revealed as one of the countless treasures which lie concealed beneath the letter of the Word of God.

Every doctrine taught by Swedenborg is based upon, drawn from, and confirmed and illustrated by the Divine Word. Those who reject the Holy Scriptures, and those who cannot be brought to see their interior glory by means of a better and higher interpretation, will probably not understand all we have to say, and must be left to feed upon the husks of naturalism, while in their Father's house there is bread enough and to spare.

The first chapter of Genesis,—rejected by science, doubted by reason, the jest of the infidel, the stumbling-block of the Christian,—contains psychological truths of the highest order; truths so transcendent and universal, that volumes might be written in elucidation of each verse as a text. From this source we have drawn the principles which underlie our theory of love, sex and marriage. Swedenborg's exposition in his Arcana Cœlestia, refers specially to the changed relation between the will and understanding occasioned by the fall; but the same principles are applicable to the relation of husband and wife, or of the Lord and his Church. Many of the most beautiful truths are even apparent in the letter of the Word.

Man, the image of God, contained the female element *within himself* when first created.

"Let us make man in our image and after our likeness."

The plural number is here used, because the Divine Love and the Divine Wisdom, or the bi-sexual elements in the Divine nature, are the creative power which gives to man the bi-sexual character of his own organization.

"So God created man in his own image; in the image of God created he him; male and female created he them."

Again says the Divine Word:

"In the day that God created man, in the image of God made He him:

"Male and female created He them; and blessed them, and called their name Adam in the day when they were created."

Adam, therefore, containing within himself both the male and female elements, was the image and likeness of God.

Why was the female element taken out of the man, so that the sexes became distinct and objective to each other?

If the love or female principle had remained within the man, exciting, attracting, and vivifying his wisdom-principle, man would have become a form of self-love, and would have been withheld from any external love, use or activity. But it was taken out of the man, and presented to him in a beautiful external form, an image or counterpart or complement of himself, so that he might love himself in another. This mutual love of *one's self in another*, or of another as one's self, is the conjugial love—the fountain and mother of all loves.

The ancients represented, by the graceful myth of Narcissus, what man would have been if the female element had not been taken out of him, and presented to him in a beautiful external form. Narcissus was a beautiful young man who died for love of his own face in the water. He had no attendant but Echo—the shadow of himself; and he consumed away because he failed to realize his ideal. If his Echo had been turned into an Eve, he would have discovered a paradise of uses, beauties, and glories all around him. After death, according to the story, he was changed into a flower, which was made sacred to the infernal powers; in regard to which Lord Bacon says: "Whatever produced no fruit in itself, but passeth and vanisheth as if it had never been, like the way of a ship in the sea, the ancients were wont to dedicate to the ghosts and the powers below."

This creation of Eve out of Adam, is also the symbol of the larger creation of the whole universe of souls out of the bosom of God—the All-Father. The Love of God, yearning for something to love outside of his own perfections, takes an objective form in the created universe, drawn from his own substance and vivified by his own life; and this form is a female form, and known in its largest and grandest signification as the Church of God, embracing all who are capable of receiving, in the greatest or least degree, the descending influx of the Divine Love and Wisdom.

Thus Christ, or the Divine Wisdom, the Divine Truth, "the Light of the World," becomes the Bridegroom, the Husband of the Church. Their love is conjugial, reciprocal, eternal.

Wisdom is only satisfied with love; love is only satisfied with wisdom. Each without the other is nothing; united they are all things. Each sees in the other itself beautified and glorified, its hope, its dream, its heaven, its all.

Their attractive or passional affinity for each other, yearning to become one in form, soul, function, use, and life, is love. This conception of love includes all loves, as the sea includes its waves.

Neither God, nor Adam his image, could be satisfied with the self-analysis and contemplative powers of the introverted understanding alone. The living soul, inspired with wisdom by love, yearns for something outside of itself to reciprocate its affection; something which it can call its own, and on which it can lavish all the activities of its being. Without such an external object, man is said to be "alone," and the proper reciprocating object is called a "help meet for him." That he might not be "alone," God created a universe of souls, a vast spiritual Church or Bride for himself; and repeating the same work in Adam, He blessed him with Eve. There is another higher and to us more obscure meaning of the word "alone." It implies a celestial state of life and marriage in which the soul is one with the Lord, and sees only Him in all things of the created universe. The descent from that highest state into a lower but still holy and beautiful spiritual state, is meant by the deep sleep falling upon Adam, and the creation of an external feminine counterpart to his interior life. The expulsion from the Garden of Eden is the descent, through disobedience, from the spiritual into the natural degree of life in which we still live.

The understanding, or properly male element, which penetrates and analyzes all things, is not satisfied with the infinite displays of God's wisdom and power and glory in the universe around us. Great and dazzling as they are, they are as cold and distant and lifeless to our hearts, as the stars of a summer night. We may know all things and all creatures, divine their properties and call them by their names, as Adam did; but the "help meet" is not found there. We can find it only in something taken out of ourselves and therefore our own; bone of our bone and flesh of our flesh; something in whose radiant beauty and life and love we can realize the unspeakable dreams and intuitions of our inmost souls.

The utter loneliness of the man whose better half of life has not yet been made objective to his senses, is thus described by the poet of Hope, with a sweetness which ever bears repetition:

> "Till Hymen brought his love-delighted hour,
> There dwelt no joy in Eden's rosy bower;
> In vain the viewless seraph, lingering there
> At starry midnight, charmed the silent air;
> In vain the wild-bird caroled on the steep,
> To hail the sun, slow wheeling from the deep;
> In vain, to soothe the solitary shade,
> Aerial notes in mingling measure played;
> The summer wind that shook the spangled tree,
> The whispering wind, the murmur of the bee:—
> Still slowly passed the melancholy day,
> And still the stranger wist not where to stray.
> The world was sad! the garden was a wild!
> And man, the hermit, sighed—till Woman smiled."

Woman thus created was not another man; nor was she a creature of a higher order; for she was created *out* of Adam and *for* him. She is a part of him; bone of his bone, flesh of his flesh; as he recognized her to be, when he called her his wife and spake those remarkable words:

"Therefore shall a man leave his father and his mother and shall cleave unto his wife: *and they shall be one flesh.*"

A charming writer, full of deep and spiritual thought, maintains that woman is a distinct or discrete creation, coming after man and superior to him; having more organs, functions, uses, and powers, and therefore destined to rule the world.* Love or woman will certainly rule and save the world; but it will be done by working in married harmony with wisdom or man, its conjugial partner.

Anatomically speaking, man and woman have the same organs; but they have different degrees of development according to the functions of paternity or maternity to be exercised. What is fully developed in one sex, is rudimentary or scarcely visible in the other. And all of woman's highest functions, like herself, are still only a part of man; for it is her greatest office to ultimate on the material plane of being the forces with which she has been impregnated by the masculine element from the spiritual side. And there lies a grand mystery.

Paternity and not maternity is the crowning phenomenon and wonder of nature. We can partially understand the maternal aspect of

* Mrs. Farnham's Woman and her Era."

vegetation; the bosom of mother earth; the nutritive elements provided by air and water; the chemical combinations which build up the new and beautiful structures before us; but the impregnating power of the solar ray, the paternal aspect of it all, remains an inscrutable mystery.

Swedenborg tells us that the soul of the child comes from the father. The body, with its animal spirits derived from the mother, would be a mere human animal, irrational and mortal as the Undine of fiction, if it were not spiritualized by the paternal forces. This is not the place to discuss the physiological or theological bearings of this great truth. We only wish to affirm that woman with her maternal functions and forces is an appendage to man, a complementary and supplementary form, through which his spiritual forces are given a physical outbirth and embodiment.

In one sense woman is superior to man. She is the central, highest figure in the creation. The last created, receiving first the Divine life, she intervenes, as it were, between man and heaven. The vegetable kingdom is formed out of the mineral, and lifts it higher toward the vital world. The animal kingdom is formed from both, and comes still nearer to the spiritual. Man is builded upon all the kingdoms below him, and connects with heaven. Woman is not another kingdom separated from him, but a part of himself, reaching higher into the celestial atmosphere; and, like the lofty places of our earth, she is first illumined by the ethereal touches of the coming Day.

It is generally supposed that with the same education, the same opportunities and pursuits, the same surroundings, etc., man and woman would be very much alike, and indeed almost undistinguishable. It is not so, however. Woman's organic instincts would finally assert themselves under any conceivable pressure of external training. She would surely revolt against any life which did not involve and effect her thorough and spiritual union with man, such a total immersion of self in the conjugial partner as to produce really *one flesh*.

She is not the equal of man; she is not his superior. She was taken out of him; she belongs to him; she is his interior self in external form. She is nothing but a shadow, or image in a mirror, without him. She yearns for union with him, as he does for union with her, more than for all things else. Neither man nor woman would have a destiny or even a life without the other. It is their union which makes the perfect Man, the image and likeness of God, just as the union of the Lord with his Church will make the perfect world and the perfect universe.

The bi-sexual nature of the Supreme Being is decomposed in the sexes as light is decomposed in the prism. Break a magnet into two pieces, and each part is still a magnet with its positive and negative or male and female poles. Man and woman stricken off, to speak analogically, from the Divine substance and life, remain both masculine and feminine in themselves, although objectively masculine and feminine to each other. Every man has a feminine, every woman a masculine, element more or less developed. *The key to the differences between the sexes is this: the poles are reversed.*

Wisdom, as revealed through the operations of the understanding, is the positive or ruling element in man, and dominates or determines the character of his love.

Love, as revealed through the activity of the affections, is the positive or ruling element in woman, and dominates or determines the character of her wisdom.

The interior life and vital principle in man lies in this, that he is inspired with the love of the divine wisdom, that is, of every external manifestation of the divine glory, constituting the whole of knowledge, science, art, philosophy, and all things upon which the understanding is exercised. The interior and vital soul of woman is not the same. It is the love of all these things, not for themselves but as they have been acquired by and exist in man. She loves man for all those things for which he would naturally love himself, and that is the reason why he is so warmly attracted to her.

Othello's wooing of Desdemona by the story of his wonderful "battles, sieges, fortunes," hairbreadth escapes and "moving accidents by flood and field," very beautifully illustrates this truth:

"My story being done,
She gave me for my pains a world of sighs:
She swore—in faith 'twas strange, 'twas passing strange,
'Twas pitiful, 'twas wondrous pitiful!
She wished she had not heard it: yet she wished
*That heaven had made her such a man.*

* * * * * * *

She loved me for the dangers I had passed,
And I loved her that she did pity them."

Milton also caught a glimpse of the spiritual relation of the sexes when he said of Adam and Eve:

"In their face
The image of their glorious Maker shone.
For contemplation he, and valor formed,
For softness she, and sweet attractive grace;
He for God only, she for God in him."

Let the reader ever bear in mind that in the following statements of the influence of the sexes on each other, we have presumed both parties to be living in true Christian order, in obedience to the commandments of God. It would be a great mistake to suppose that all these beautiful things are asserted of men and women in the present corrupt and disorderly conditions of society. They are the possibilities of the race.

Man's love is abstract; woman's concrete. His love is for principles; hers is for persons. Man loves himself in woman, but woman loves man unselfishly Hence the intensity, the individuality, the singleness, the spirituality or super-sensuousness of her love. Every woman is the organic form or realization of some man's intellect, loyal only to him and his, and yearning and groping through the great dark of our natural life for a perfect and eternal union with her conjugial partner.

Eve's intuitive perception that man's life was his love of wisdom, and that her own wisdom was due to her love of man's wisdom, led her to give a willing ear to the temptation of the serpent, or sensual principle, which said:

"In the day ye eat thereof, then *your eyes shall be* OPENED; and ye shall be as gods, *knowing* good and evil.

"And when the woman saw that the tree was good for food, and that it was pleasant to the eyes, and a tree to be desired *to make one wise*, she took of the fruit thereof and did eat; and gave also unto her husband with her, and he did eat.

"And the eyes of them both were opened."

Woman thus sensualized both herself and man to gratify him, as she is doing at the present day. When man acquires interior and heavenly wisdom, woman will inspire him with a love entirely super-sensuous and celestial.

The zealous devotion of woman to church and state, and to all the institutions established by man, is simply the expression of her love of the masculine wisdom by which all these things have been achieved or constructed. She is conservative and aristocratic where he is; republican and radical where he is. She attaches herself to him and his, and identifies herself with them. The women of the Southern States cared little or nothing for the principles which animated the Southern soldier; nothing for slavery or separate nationality. Their self-sacrifice, which has never been exceeded in the history of the world, was made for persons; for husbands and sons, for lovers and brothers. Thus was it always; thus will it always be.

There is no height to which woman cannot rise, no depth to which she will not fall, in obedience to the great law of her being, that she must ultimate in external forms of beauty, love, and use, the secret life with which she is impregnated by the masculine sphere of thought. The holy women who accompanied Jesus, intuitively divining the spiritual truths of his kingdom, outstripped in faith and devotion the tardier understanding of the apostles. Vivified by the same truth, it was a woman who taught with practical sublimity the doctrine of the universal brotherhood of man, long before men had more than dreamed of giving it political and social expression. The story is this:

"St. Catherine of Siena on one occasion accompanied a notorious malefactor to the scaffold, and administered to him the consolations of religion. When the axe had done its duty, she lifted up the bleeding head, and addressing the assembled crowd by whom she was regarded as a direct messenger of God, she said: 'Fear not to raise your prayers for this man's soul. I accept him as my brother;' and she kissed the dead lips."

Swedenborg says that with man love is interior and wisdom exterior, while with woman wisdom is interior and love is exterior. Hence the strength and power of man's external form and the grace and beauty of woman's. Hence also man is the creator, the father, the builder, the organizer; for the divine energies work through his interior life and manifest themselves by means of his understanding. Woman receives not the divine wisdom, but the male wisdom or understanding, into her interior life, and there it becomes intuition, a spiritual instinct; and she brings it all down into nature again, and gives him back the True clothed in the garment of the Beautiful. Unconsciously to herself and to him, she transfigures the crude shapes of his earthly thought into apparitions of spiritual loveliness, which excite and purify his affections for the reception of still higher degrees and forms of wisdom.

Of the grandest phenomena in nature our poor spirits are wholly unconscious. We feel nothing of the transcendent miracles which are being wrought every moment in our own bodies. The weight of the atmosphere, the whirling motion of the earth on its axis, the magnetic currents encircling the globe, and a thousand other wonders, are all imperceptible to us. We know nothing, also, of the spiritual currents which proceed to and fro between the sexes, binding them together for the uses of life, as the heart and lungs are bound together in the bosom of the living man.

Love and Wisdom are the opposite poles of the magnet, which always attract, sustain, excite and intensify each other. Man's interior life being the love of divine wisdom, is impregnated by the Divine Wisdom, and the result is the masculine form and understanding. Woman s interior life being the wisdom or understanding of the man, is impregnated by the Divine Love, and the result is the ineffable charm and beauty of the female form. Such is the influx of the Divine life into both sexes. Thence follows the operation, by spiritual influx, of the sexes on each other.

One current starts from the Divine Love as a centre, and flows through woman; for woman is receptive of the divine feminine, and ultimates the love she receives in beautiful forms and uses. The other current starts from the Divine Wisdom as a centre, and flows through man who is receptive of the divine masculine, and ultimates the wisdom he receives in the manifold organizations or institutions which are the product of thought. The love manifested in the beauty and goodness of woman, is the attractive centre to the wisdom which is the life of the masculine understanding; and *vice versâ*. This is the spiritual origin of sex, love and marriage, which, in their largest sense, are of necessity universal and eternal.

Woman inspires man with love, because she is the very form of his self-love or the love of his own wisdom. Man inspires woman with wisdom, because his wisdom is the magnet which excites the activity of her love.

She impregnates him with love which quickens his rationality, and ultimates itself through the intellectual faculties.

He impregnates her with wisdom which quickens her emotional intuitions and perceptions, and ultimates itself through the affections.

> "Her rapid mind decides while his debates;
> She feels a truth which he but calculates."

Woman absorbs the masculine forces and brings them out into the material plane of life. Man responds to the excitation of the feminine principle by ultimating its forces on the spiritual or mental plane of life. Thus is the great spiritual current from heaven to earth and back again to heaven, established by means of the co-operating influence of the sexes. Either sex without the other, would be as useless and lifeless as one half of the human body separated from the other. Conjoined in heavenly marriage, they constitute the man or the angel.

Patmore alludes to this mystical and reciprocal mental action of the sexes on each other:

> "Love kissed by wisdom wakes twice love,
> And wisdom is through loving wise."

That woman stimulates the intellectual activity of man by the love with which she inspires him, is written on every page of the public and private history of mankind. Man's wisdom comes immediately from God; woman's wisdom comes mediately through man. Woman's love comes immediately from God; man's love comes mediately through woman. As love is the centre and life of all things, and as wisdom or truth is determined by the love from which it springs, woman is the *motor power*, while man is the *organizing power*, of human life. Woman is the heart, man the head, of the social body: woman is the priest of the world, if man is its king.

The grand psychological truth involved in the above paragraph, is sweetly recognized by a charming English poet:

> "For Love is substance, Truth the form;
> Truth without Love were less than naught;
> But blindest Love is sweet and warm,
> And full of Truth not shaped by thought."

All love, therefore, comes from woman: all thought comes from man. Thought flowing from the man into the woman, becomes feeling. Feeling flowing from the woman into the man, becomes thought. This idea is well illustrated by Dryden in his poem of Cymon and Iphigenia. Cymon was a poor stupid creature whom no pains or labor could instruct or improve. He one day found the beautiful Iphigenia sleeping in the woods, and a ray of love's divine light illumined his soul, causing his torpid understanding to spring into sudden activity. The story is a true symbol of a universal law.

"Through the rude chaos thus the running light
Shot the first ray that pierced the native night:
So reason in his brutal soul began;
Love made him first suspect he was a Man;
Love made him doubt his broad barbarian sound;
By Love his want of words and wit he found:
That sense of want prepared the future way
To knowledge, and disclosed the promise of a day.
What not his father's care nor tutor's art
Could plant with pains in his untutored heart,
That best instructor, Love, at once inspired,
As barren grounds to fruitfulness are fired:
Love taught him shame, and shame with Love at strife,
Soon taught the sweet civilities of life."

Yes; man has no love, no passion, no sentiment which does not primarily come to him from woman. Woman alone has the power of drawing the divine fire down from heaven. The mother, the first love, the wife, are the sources of our purest and best inspirations and wisest thoughts. And from the whole female sex a vast current of affection pours toward the whole male sex, inspiring it to intellectual activity, like the mighty ocean stream which bears the tropical heats in its bosom, and causes the bleakest shores of the North to brighten and blossom with a life not their own.

In the wildest mining districts of California, and in communities of men utterly lawless and desperate, the advent of a single American woman, unknown to nearly all of them, was a signal for the decrease of violence and the dawn of peace and order and law; while visions of home and happiness and flower-gardens and children, of books and music and love, of schools and churches and heaven, arose in every mind.

"I believe if I had something from the hands of a woman, I would get well," said a dying English soldier in a hospital of the Crimea; and another brave fellow, who got hold of the hand of one of Florence Nightingale's nurses, grasped it tightly in silence till he died!

The unhappy and licentious Byron, yearning on his deathbed for all that remained to him of the pure and the good in woman, exclaimed with his last breath:

"Augusta! Ada!—my sister! my child!"

As woman alone draws love down from heaven to warm the world, so man alone draws down wisdom for its light and guidance. What then is the understanding of woman?

A reflection of the masculine mind, just as the light of the moon is the reflected light of the sun. Milton speaks wisely of the sun and moon as

> "Communicating male and female light."

The intellect of woman is soft and tender, differing from that of man as the sweet face of woman does from the sterner features of the male sex. Woman, by the intensity and purity of her love for man, is capable of absorbing the whole of his wisdom, so that she shall appear fully equal to him in all the labors of the understanding. She may reach the higher masculine standard in astronomy, mathematics, science, or literature. Still, her intellectual light is borrowed. She originates nothing, not even in the composition of music, in which her organization would lead us to think she would excel. The office of her intellect, indeed, is not to originate or organize, but to utilize or make fruitful in her own field the stores of wisdom which man has acquired in his. And here we detect how the intellect of woman is the greatest and best "help meet for him" that he can possibly find. Educate woman up to the masculine standard of thought, and fire her soul with the love of God and the neighbor, of husband and children, of home and country, and the world will find in the expression of woman's opinion on every subject a new fountain and oracle of true wisdom hitherto unknown.

When man and woman love purely and are truly conjoined, the subtlest combinations of the male intellect are sometimes futile and weak in comparison with the lightning-like wisdom which emanates from the feminine brain,—a fact, which may be of immense importance to the future of human society. This psychological truth is beautifully illustrated by Schiller in the character of Thekla in his Wallenstein.

When Wallenstein and his sister are weaving the subtle web of treason about the young Piccolomini, whose frank and brave nature is as incapable of suspicion as of fear, Thekla, his betrothed, visits the camp. Her intuitions, quickened by love, immediately perceive the danger his intellect has not detected; and the moment they are alone she rushes to him, exclaiming,

"Do not trust them! They are false!"

When the fatal step is taken, the exposure made, and Wallenstein is an open traitor to his emperor, Piccolomini still lingers in the camp, magnetically attracted by his love for Wallenstein's beautiful daughter. All but Thekla endeavor to draw him over to the revolt. They ply him with persuasions and powerful arguments. His reason staggers, blinded by love. Suddenly he appeals to Thekla to think for him, to decide for him. He leaves the whole matter "to the unerring good within her heart." To the rage of her friends and the despair of her lover, she sublimely sacrifices their hopes and interests, and counsels him to the path of duty and honor:

"Being faithful
To thine own self, thou art faithful, too, to me.
If our fates part, our souls remain united.
A bloody hatred may divide forever
Our houses, Piccolomini and Friedland,
But *we* belong not to our houses. Go!"

It may be asked, How is man the organ or

medium of the divine wisdom, if his own wisdom is determined by the inspiring love of woman? In spiritual and natural things alike, conditions determine development. Sunlight cannot create flowers or fruit until seeds are properly planted. Thought is impossible until the affections are awakened. The distinctive peculiarity of the male mind is, that man's affection, awakened by woman's influence, is impregnated by the inflowing divine wisdom, producing all the resplendent phenomena of the intellectual or rational sphere. The love of man's achievements is the heart-life of woman; and that, impregnated by the Divine Love, is the source of her intuitive wisdom, her beauty, and her grand capacity for offices of self-sacrifice, love, and use.

In the presence of these fundamental laws of our spiritual being revealed by Swedenborg, it is entirely useless to discuss the common question of the equality or inequality of the mental and moral powers of the sexes. Man is at present inferior to woman on the moral and woman is inferior to man on the intellectual plane, simply because the sexes, as the two great halves of our spiritual life, are not conjoined in the heavenly marriage. Woman is capable of absorbing and reflecting the entire understanding of man, so that her intellectual faculties shall be identical with his. Man is capable of rising to the loftiest height of woman's love. Nothing, however, but the universality of marriage can accomplish such a glorious equality, and that kind of marriage in which each sex receives in perfect measure all the other has to give, so that they become, in the no longer mystical words of the Bible, *one flesh.*

This mutual, organic, inevitable, beautiful dependence of the sexes on each other for the manifestation of their peculiar properties, making co-operation and marriage necessary, here and hereafter, overturns the prevalent theories of sex and marriage, and leads to a higher philosophy which will solve the mysteries of beauty and love and define the sphere and duty of both man and woman.

From the universality of sex, or the bi-sexual character of all the forms in nature, and from the fact that spiritual truths are always pictured in natural laws, we may arrive at the following important conclusions in our pursuit of truth:

The excitation of one species of electricity is always accompanied by the excitation of the other, both being produced in equal degrees. This is true also of love and wisdom, or affection and thought, the spiritual electricities of our life. Given the state of wisdom in man, and you have the exact measure of the state of love in woman; and *vice versâ*. Every augmentation of woman's love increases the wisdom of man; which suggests the true sphere of woman, and the true method of moving, reforming, and spiritualizing the world.

As the strength of the magnet depends upon the intensity of the two opposite poles, so the perfection of human character depends upon the distinct rationality or masculinity of man, and the distinct emotional activity or femininity of woman.

Like the two poles of a magnetic bar, man and woman are the mutual sustainers of each other's condition; he of her love, she of his wisdom. Man grows more truly rational and masculine, as woman grows more truly feminine.

If man grows more effeminate and woman more masculine, they both lose the proper influence they should exert on each other, because each sexual pole of life is weakened by the other.

The nearer they approach each other in quality and pursuit, not by reflecting but by rivaling each other, the less will be the reciprocal attraction between them.

If the distinctive spiritual characteristics of the sexes could be destroyed, they would cease to respect the chastity or honor of each other; spirituality of thought and motive would become impossible, and man would sink to the level of the brute.

These are great and eternal truths, which might be scientifically demonstrated and confirmed by countless illustrations from human experience and history. Let them serve as beacon-lights to those earnest and daring souls who are striving to reconstruct the social fabric upon principles derived from the supposed light of nature, and not always illumined by the superior light of spiritual truth.

The inspiration of the poet is ever in harmony with the wisdom of the Seer:

> "For woman is not undeveloped man,
> But diverse: could we make her as the man,
> Sweet Love were slain: his dearest bond is this,
> Not like to like, but like in difference:
> Yet in the long years liker must they grow."

If man and woman were equal and alike; if their wills and understandings had no differences but such as exist between parties of the same sex; then the spiritual marriage or absolute union of two souls would be impossible. No such union can exist between two men, or between two women. It is only possible between the sexes, because they are unequal and unlike; because man derives his wisdom from God and his love mediately from woman, while woman derives her love from God and her wisdom mediately from man. This is the grandest psychological truth ever revealed to the world.

All life flows from the spiritual world into the natural. All love, all wisdom, all our affections and thoughts, come down from the interior. Woman is the organ or medium for the descent of the divine love. Man is the organ or medium for the descent of the divine wisdom. The love which woman acquires from above, is reflected over to man. The wisdom which man by means of that love acquires from above, is reflected over to woman. The correspondence and attraction between the love and wisdom thus acquired by the co-operation of the two sexes, is the basis of the spiritual or heavenly marriage.

Now how is the actual amount of love in the world increased? or what is the special office of woman? And how is the actual amount of wisdom in the world increased? or what is the special office of man?

As there is no wisdom without love, so there is no love without obedience. Obedience, or living according to the divine law, is the means whereby love and wisdom are brought down to us. We become truly wise only by keeping the commandments. The mind cannot be permanently illumined until the heart is purified. Obedience was properly made the test of the possibility of such a spiritual life as that signified by the Garden of Eden. Disobedience brought sensuality, blindness, spiritual death.

It is the obedience of woman to the divine law, which brings down the heaven of love to the world. It is the obedience of man to the divine law, which brings down the heaven of wisdom to the world. Thus the work of each sex is complementary to that of the other. It is only when they co-operate or act in harmony, that true love and true wisdom can be obtained.

The sphere of man's labor is evidently the sphere of the understanding, beginning from without and working upward and inward. The sphere of woman's labor is the sphere of the affections, beginning within and working downward and outward. Woman's love is supersensuous; man's love is sensuous until spiritualized by woman. Man's understanding is rational and inductive; woman's understanding is intuitive and deductive, because it is only man's understanding raised to a higher or spiritual power. Man's forces are centripetal, woman's are centrifugal. The masculine ascends the ladder of Jacob to heaven, but the feminine descends it with blessings for mankind.

Man's labor is in the field of the senses. He examines, compares, discovers, invents. He conquers nature, and the material is to him the real. He tills, plants, reaps, builds, travels, creates. Science, literature, art, philosophy, manufactures, governments, churches, institutions, are the products of his strength and wisdom, the fruits of his organizing mind; and all these he lays at the feet of woman—the silent, secret, unseen motor power of all his goodness and wisdom and courage and labor.

Woman works in a far more interior sphere. To her sweet intuitions the spiritual is real and supereminent. She alone saw the angels at the sepulchre of the Lord. She repays man for his benefits by bringing down into nature the ideal light and beauty of an inner world. She conquers man, not nature; and her weapon is not the strength of reason, but the inspiration of love. She inspires him with that tender sentiment which stimulates him to new thought, new wisdom, new power. She etherealizes and spiritualizes all about him. She blesses him, elevates him, purifies him. She gives him love which is her dower from God, and home which is the image of heaven.

The words *wife* and *mother* express the

whole nature, functions, uses, and life of woman. No education, training, business, or influences, which do not help to qualify woman for these sublime offices, are of any permanent value. These words, wife and mother, so full of tenderness and beauty, have a spiritual as well as a natural meaning. Nature designed that every woman should be a wife and mother in both senses; but the evils of our imperfect civilization leave many a virgin rose to wither on the thorn. Those women are spiritually wives and mothers, whether externally and legally married or not, who absorb from the masculine element of thought in books, in conversation, in society, in practical life, in religion, the subtile forces of wisdom which enrich their own souls with grand intuitions, beautify their characters and faces with wifely and maternal purity, modesty, and humility, and produce around them a crowd of radiant influences, sympathies, divinations of love, and unconscious charities, which are the children of the spiritual marriage.

Woman's work is within. What is external in man is internal in woman. Organically, mentally, socially, spiritually, she is more interior than man. She herself is an interior part of man, and her love and life are always something interior and almost incomprehensible to him. He never understands woman as woman understands him. The man's thought reverts occasionally, or it may be frequently, to wife and child. The woman's thought stands always like an invisible angel by the side of husband and child wherever they may be. She is the unseen Ariel who creates all the sweet and mysterious music which leads us out of the dark and thorny places of life. "O woman!" exclaims Michelet, "fragile globe of incomparable alabaster, wherein burns the lamp of God."

In the old world and the old times, when everything was symbolic, this *interiority* of woman was represented by the large mantle in which the bride was concealed when presented to the husband. It was by this means that Jacob was deceived, and Leah substituted for Rebecca. The modern veil is the last vestige of this concealment. The strict seclusion of the Oriental woman, the flowing, concealing dress of the sex, the scriptural command that woman should not go with the head uncovered, nor speak in public assemblies, have their origin in a spiritual recognition of the essential interiority of woman.

For the same reason the sphere of woman's power is the house, the chamber, the closet, and not the street, the field, the forum, the market-place,—the common theatres of masculine activity. Woman is to deal with domestic affections and uses, not with philosophies and sciences. She is the form of charity, not of faith. She is priest, not king. The house, the chamber, the closet, are the centres of all social life and power, as surely as the sun is the centre of the solar system. Woman occupies already the precise point from which she can move the world. She needs only the lever, which is the knowledge of the right method of doing it.

Another proof of the interiority of woman, is the wonderful secretiveness and power of dissimulation which she possesses. These words applied to woman's character, have a good meaning very different from the usual meaning attached to them. Woman's secresy is not cunning; her dissimulation is not fraud. They are intuitions or spiritual perceptions, full of tact and wisdom, leading her to conceal or reveal, to speak or be silent, to do or not to do, exactly at the right time and in the right place. A type of woman's beautiful dissimulation is seen in the mother of Moses, who carried her point with great ingenuity, concealing her child from his cruel enemies; and, what was still more difficult, concealing her own gushing joyous maternity from the Egyptian princess.

A charming English poet, describing the secret and wonderful power of a true woman to elevate and purify the man she loves, thus sings:

"Without his knowledge he was won,
Against his nature kept devout;
She'll never tell him how 'twas done,
And he will never find it out.
If, sudden, he suspects her wiles,
And hears her forging chain and trap,
And looks; she sits in simple smiles,
Her two hands lying in her lap!"

Such are the spiritual differences between man and woman, on which must be based a true philosophy of sex. Eminent service will these doctrines render to those who study the spiritual uses of marriage, the causes and effects of the perversion of the love-principle or of false marriages, and the means whereby may be inaugurated a new and better civilization, in which woman, as love married to wisdom, or charity conjoined with faith, shall share with man the government, the glory, and the happiness of "a new earth," fitted by their united efforts for communication with the "new heaven."

In all we have said thus far, we have had only good men and women in our mind's eye. There is a dark reverse to our bright picture; opposite forms, but governed by the same laws. Instead of good, there is evil; instead of the true, there is the untrue; instead of heat, cold; instead of light, darkness. There are men who breathe upon women only the sensual and the false. There are women "whose chambers are the gates of death, and whose steps lead down to hell."

## CHAPTER VI.

### *THE SPIRITUAL PHILOSOPHY OF LOVE AND BEAUTY.*

O YOUTH! beautiful, tender, enchanting aurora of life! happy transition period between the child and the man! What soul from whose hearing

"The horns of Elfland, faintly blowing,"

have not died away for ever, can forget its sweet illusions, its wild ambitions, its incommunicable longings, its transport and its tears?

"Tears from the depths of some divine despair!"

The wonderful secret of all this, is love. God created woman to be a form of love. At the age of puberty the magnetic forces of the sexes begin reciprocally to excite and attract each other. Up to that period, the boy and girl are almost undistinguishable in form and feature; then they separate,—separate, however, only to be drawn together again by irresistible powers, and reunited for ever.

The astonishing changes which take place in the physical and moral systems of both sexes at this period, are explained by the doctrines advanced in the preceding chapter. Like flowers which have different affinities for the heat and light of the sun, the feminine life awakens into activity at the touch of the Divine Love, and the masculine life at the touch of the Divine Wisdom. The feminine principle, secretly inspired by the tenderest love, clothes itself with beauty in which the corresponding masculine intellect can see its own interior wisdom portrayed and reflected as in a magical mirror.

Therefore the phenomena of woman's development which differentiate her from man, lie in the plane of love or the emotional sphere; and the phenomena of man's development which differentiate him from woman, lie in the plane of the understanding or the intellectual sphere.

This law is represented in the body of each sex by specific changes. The changes in woman belong to the interior, those in man, to the exterior, organic forms. In woman they attach to the blood-system, of which the heart, the great representative of the emotional life, is the centre. In man they attach to the respiratory system, for the lungs reflect and ultimate by speech the intellectual life of the brain. The interior periodic changes in woman, announce her preparation for a love more spiritual and super-sensuous than that of man, and for the holy offices of wifehood and maternity. The change of voice in man and the growth of muscle and beard, proclaim the development of wisdom, courage, power, and "the wrestling thews that throw the world."

Inspired by the love of wisdom or knowledge, the masculine understanding early asserts itself and betrays its spiritual affinities. What earnest pursuit, what abstraction, what passionate study, what lofty ambition, what hope, what audacity! Imagination lends wings to desire, and the dreams of the youth are prophecies of the man. The young statesman declaims to applauding senates which have never assembled, and the poet who has not yet sung, listens to his songs as they echo round the world. The coming soldier steps to inaudible drums, and the born sailor hears in his inland solitude the music of the sea.

How different the evolution of woman's nature! The bloom of a new loveliness comes over her face. The charm of a new elegance like a spiritual drapery envelops her form. Startled at her own dawning beauties which make her the centre of attraction, she retires, self-conscious, into herself with bashful timidity. She trembles, "like a guilty thing surprised," at the unfoldings of nature in herself and others. What charming apprehensions, what grace, what sweet dependence, what affection, what passion, what caprices, what tender inquietudes, what brooding fancies!

> —"And hopes, and fears that kindle hope,
> An undistinguishable throng!"

—all foreshadowing the development of a moral and emotional medium or organ of the Divine Love; the bride of wisdom, the rose of joy, the pearl of innocence, the light, the life, the wonder of the world.

Then it is that the two angels of youth,

Honor and Chastity, male and female, appear with their flaming swords to guard the sacred gates which enclose the tree of life. Chastity is the purity and innocence of love, and modesty is her inevitable handmaid. Honor is loyalty to truth and principle, and courage is his noble shield-bearer. Chastity belongs to the heart and the will; Honor to the understanding and the thought. The gender of these virtues is the reason why we speak of the honor of man and the chastity of woman as the priceless pearls of a high civilization.

Love radiating from its divine source, changes its manifestation according to the forms into which it flows. In the highest or celestial sphere, nearest to the Lord, it is the celestial love of the angels. In the next inferior sphere, it is spiritual love. Lower still, it is the natural love of the sex with man. Below the rational sphere, it is simply animal or corporeal. Divested of sentiment, it becomes a mere passion. Farther down the ladder of nature, it is the sympathy of plants, the attraction of metals, the gravitation of atoms,—love being the life and soul of all things from the greatest to the least.

Woman is a bundle of loves which determine her thought. Man is a bundle of thoughts which determine his love. As every chemical element in nature has specific affinities for certain other elements, so every constituent of the masculine soul has its fixed affinity for some corresponding constituent in the feminine soul. The elemental forms of our interior minds are forever striving after union with their sexual counterparts. This is the origin of the love of the sex, or the mutual and reciprocal passional attraction between man and woman. Thence come all the activities of life.

Woman is attracted to man by the manifestation of those forms and qualities which flow from the operation of wisdom or truth in the masculine soul. A woman may marry from any motive,—external, inadequate, and even base; but she never loves a man unless he is wise, brave, strong, just, or capable, in her eyes. She does not care for beauty in him; because she wishes that for herself, so that she can bestow it upon him in exchange for the protection, the power, the dignity, the organizing and executive genius of the masculine life.

Man is attracted to woman by that beauty which is the symbol or hieroglyph of his own most secret thought. Her grace of movement, symmetry, softness and elegance of form, her sweetness of manner, vivacity of intellect, or tenderness of sentiment, strike the hidden and sympathetic cords in some masculine soul, which thus sees its counterpart or other self revealed to it in a manner incomprehensible to all other souls.

> "He meets, by heavenly chance express,
> His destined wife: some hidden hand
> Unveils to him that loveliness
> Which others cannot understand."

Each masculine or feminine soul has its organic peculiarities which differentiate it from all others, as each face differs from all other faces. Add to this, that each soul finds its corresponding complement in the other sex, and we see how the love of the sex has a constant tendency to terminate and resolve itself into a love of but one of the sex. This is the natural cause of marriage as an institution, which, by civil and religious sanction, enables us to secure and retain sole possession of the object of our love.

"Marriage," says Jeremy Taylor, "is the mother of the world; and preserves kingdoms and fills cities and churches, and even heaven itself."

Of this reciprocal influence of the sexes, this tender passion, the sweetest bond of social life and the inspiration of poetry and art, Emerson beautifully observes:

"A private and tender relation of one to one is the enchantment of human life, which, like a certain divine rage or enthusiasm, seizes on man at one period, and works a revolution in mind and body; unites him to his race; pledges him to the domestic and civic relations; carries him with new sympathy into nature; enhances the power of the senses; opens the imagination; adds to his character heroic and sacred attributes; establishes marriage, and gives permanence to human society."

It is needless to repeat what has been said by brilliant writers on the nature of love. Everything hitherto advanced on the subject, relates only to the passional attraction between man and woman in the natural life. Neither poets nor divines have obtained more than a glimpse of that spiritual love between the sexes, which is the theme and life of Swedenborg's philosophy. Its quality, even its existence, is unknown. It is claimed for it that it was revealed from heaven; and certainly it has no counterpart in the speculations of any mundane philosopher. It is the key to the doctrine of marriage in heaven. It is so novel, transcending so far the usual level of human study and thought, that it has been superficially examined, and greatly misrepresented. It deserves a thorough consideration.

Swedenborg, raising love to a higher power or spiritual degree, discards the word *conjugal*, which implies being yoked together, and substitutes *conjugial*, which means united. The conjugial love of Swedenborg is something more than the natural love between husband and wife. No purity, fidelity, or intensity of affection, can of itself elevate the natural love of the sex to the spiritual or conjugial love which is the bond of wedded souls in heaven. The most wicked men and women are sometimes capable of the most romantic passion, the most extreme devotion to the object of their love. And Swedenborg declares that the deadliest hatred may lie concealed underneath such affectionate exteriors, beyond the consciousness of both parties, only to be revealed by death and the judgment.

Conjugial love does not grow out of the natural love of the sexes for each other, but may be inserted into it like a gem in its matrix, or the soul in its body. Man is created an image of God, a finite form of heaven, and contains all the degrees of life in his own organization. He is born, however, only into the lowest or sensuous and corporeal degree. The higher degrees,—the rational, spiritual, and celestial,—may be successively opened in him by a life according to the commandments, which makes him receptive of the life of heaven. So long as he lives in the sensuous degree, he is only capable of sensual love. When he becomes rational, his love is imbued with rational life. Only as he has the spiritual or celestial degrees opened within him, can he be animated by spiritual or celestial love.

The love of the sex belongs to the external or natural man. Conjugial love belongs to the internal or spiritual man. Only the regenerate,

—only the angels understand or fully enjoy the conjugial love; and man enters into a perception of its delights, only as he advances in regeneration and becomes more and more like the angels. The spiritual degree is opened by a life of good works in obedience to the Divine Word. Just as that degree is opened, is the conjugial principle developed in the heart, whether the person be externally married or not; and the soul is becoming fitted for its eternal marriage with its mate in some angelic society.

A regenerate soul is one who has experienced the heavenly marriage between his own purified will and enlightened understanding, so that faith and charity are perfectly wedded in his nature for the production of good works. Such a regenerate soul, when masculine, is a certain finite image of the Divine Wisdom. Such a regenerate soul, if feminine, is a certain image of the Divine Love. The union between the Divine Love and the Divine Wisdom, which makes the Father and Son one Divine Being, flowing into these regenerate souls, draws them into a spiritual union,—the two together making one holy angel, or image and likeness of God.

Such is the conjugial love which makes not only possible, but probable and rational, the wonderful descriptions of married pairs whom Swedenborg asserts that he saw in heaven.

In our feeble and imperfect state, so remote from the realization of such glorious ideals, we can scarcely form an adequate conception of the conjugial love. Swedenborg declares it to be the parent and fountain of all other loves both spiritual and natural. It is full of innocence, wisdom, peace, and blessedness; and its exquisite delights are so superior to anything that springs from a merely natural relation, that they are inconceivable to the sensual man.

It is taught by Swedenborg that woman alone derives the conjugial love and the love of the sex from the Lord, and imparts them to man. It seems to man that his affections arise spontaneously in himself, and to woman that her thoughts or ideas are unquestionably her own. But it is not so. If the whole masculine sphere were withdrawn from woman, she would be incapable of thinking at all. If the whole feminine sphere were withdrawn from man, he would have no emotional life whatever. Swedenborg says the latter experiment was tried with some spirits in the spiritual world, and the result convinced them that woman is the organ or medium through whom the Divine Love is brought down to the various degrees of life.

The love with which man is first inspired by woman, has something in it especially tender and charming. We attribute the romantic sentiment of early love to the ardent imagination of youth, unchastened as yet by the severe discipline of life. It comes rather from the influx of angelic spheres redolent with the conjugial life of heaven. Celestial angels are very near us in infancy, and spiritual angels in youth; and when woman inspires us with our first love, our guardian friends endeavor to implant the conjugial principle in the embryonic forms of sexual attachment. Hence the ineffable charm and freshness and beauty and glory of "love's young dream!"

By a selfish, worldly, and ambitious life, unblessed by noble resolves or spiritual aspirations, we close the interior of our minds to heavenly influences, and our subsequent love of woman becomes more and more external and sensual. At last the cold and selfish naturalist scoffs at the idea of a purely spiritual love of woman, as a theological fiction or a poet's dream. In proportion, however, as our love for woman retains a chaste, tender and reverential spirit, do we preserve the conjugial principle in our hearts—that shekinah revealing the Divine presence,—and retain the possibility of becoming like the angels of God, who have achieved the spiritual resurrection and attained to the heavenly marriage.

This tender, unselfish, super-sensuous attraction between two souls, enriching life and defying death, is the true natural form, matrix, or cradle in which the conjugial love is best developed. It is the thorough sympathy with one, which makes us bountiful to all. "The one beautiful soul," says Emerson, "is the only door through which we enter into the society of all true and pure souls." Unions based on sensual, social, or civil grounds, may dissolve or perish. But those natural loves which emulate by their purity and beauty the conjugial or heavenly type, we instinctively pronounce immortal. Who can doubt that Abelard and Eloise found no isolating cloister in heaven? That Petrarch and Laura are no longer separated by the absurdities of rank or fortune? That Dante is at last united with Beatrice in a life more beautiful than his Poem? And that in every heavenly choir there is a loving Margaret praying for some earthly Faust?

The young heart which has been baptized with the dew of a chaste and sweet love, and which keeps itself afterward unspotted from the selfishness and sensuality of the world, can never grow old. It is not time but sin which shuts the gate of heaven. It is a beautiful thing, and truly prophetic of our immortal bloom, that man may return in his old age into the most tender and charming sentiments of his innocent youth. With what sweet simplicity and candor does the pious and learned Perthes describe the effect of a second marriage upon him in his old age!

"My own experiences amaze me. The varying moods familiar to the innocent heart of the boy in his first love; the enthusiastic tenderness which found vent in happy melancholy and universal good-will to all creation; these lay far, far behind me like a lovely dream, and no wishing had power to call them back. But now I feel again as I did then. How is this possible in a man of my age? How can I, whose heart has been so tempest-tossed by time and the world;—how can I, who have known so much and sinned so much, return thus to the innocent fondness which nestles in the newly-awakened heart of a boy?"

Poor, old, pure-hearted man! He did not know that nothing but the thin veil of death concealed from him a land of eternal youth and eternal love!

The spiritual philosophy of sex reveals also the origin and signification of Beauty,—that mysterious and ethereal power which wakens the love of man for woman into activity.

Beauty is an assemblage of properties in the form of a person or object, the contemplation of which inspires the soul with delight. Those properties depend upon symmetry, grace, color, expression, harmony, etc. Beauty is the result of the marriage union between the Divine Goodness and the Divine Truth, as manifested in human forms or natural objects. The Divine Love created the world by means of the Divine Wisdom, and the universe was therefore called by the ancients the Cosmos, or essential beauty.

Heat and light are the ultimate or last forms of the divine love and wisdom flowing into nature. The beauty of every substance is determined by the heat and light which give it organization. Its form is determined by the relation of its atoms to heat, and it is revealed to our perceptions by means of light. It absorbs, reflects, or transmits the rays of light according to its atomic structure, determined by its relation to heat. The beauty of the whole external world depends upon form and color; and form and color depend upon the relation which the atoms of all objects bear to heat and light.

This natural truth is the exact outward expression of a similar spiritual truth. What we may call the atomic form or will-principle of the soul, is determined by its relation to the Divine Love, or spiritual heat; while its intellectual manifestation or understanding, is determined by its relation to the Divine Wisdom, or spiritual light. When the two influent divine forces are perfectly married or united in the spiritual form, we have perfect symmetry, grace, order, and beauty, in proportion to the use of the form.

The life of the Divine Love manifesting itself through the Divine Wisdom, is the cause of Beauty. Therefore the Beautiful is a revelation of the True containing the Good.

Professor Oersted of Copenhagen, has given these ideas scientific expression in several æsthetic papers. He maintains that the beauty of a thing depends upon the number, character, or influence of the ideas or truths involved in it. Why do we feel that a curved line is more beautiful than a straight one? Why is a circle more beautiful than a square, or a spiral than a circle? Because the higher and more complicated form involves more great mathematical and fundamental truths than the lower, and conveys a correspondingly greater sense of pleasure to our perceptive faculties. Why is a crystal more beautiful than a common stone? Because it expresses and symbolizes more ideas and treasures of spiritual truth. Could we grasp the ideas represented by different objects, we should have the key to all of nature's symbolisms, and might immediately discover why a flower is more beautiful than a pebble, why spring is more charming than winter, why the ocean touches us with such sublime sentiments, and why we look away through and beyond the golden peace of an evening sky, into the purity, love, and infinitude of heaven!

It is not at all necessary to the sensation of the beautiful, that the understanding should have a conscious perception of the ideas or truths involved in the beautiful object. The influence of beauty is like that of music, producing a state of the soul which cannot be described in words. For the perception of the beautiful, a certain open state of the heart is required rather than an exalted activity of the head; for the beautiful awakens first the affections and then the thoughts. The poet sees more with his heart in the flowers, the rivulets, the clouds, the sunset, than the philosopher with all his erudition. And it is only the pure in heart who see God.

Reverting, now, to the spiritual differences between the sexes, we discover the wonderful office of woman in revealing the beautiful to man, and thereby uniting him with herself. The Divine Love comes to man not directly from the Lord, but mediately through woman. The Divine Wisdom comes to woman not directly from the Lord, but mediately through man. It is therefore the man's wisdom, or intellectual sphere, which the woman appropriates and marries to her own love derived from the Lord. The beauty of woman is the result of this union. Man is therefore attracted by woman's beauty, because, although he is intellectually unconscious of it, that beauty is the outward figure, symbol, and revelation of something inherent in his own intellectual organization. It is himself in another form. It is bone of his bone and flesh of his flesh. Woman is attracted to man, because he realizes for her, in his deeds of strength and wisdom, the ideals of her love; but man sees himself reflected in woman, and his wisdom mirrored back to him in the engaging forms of beauty.

The logical issue of these psychological laws is, that the beauty of woman will always and everywhere depend upon the impression made upon her by the intellectual sphere of man. All women cannot be beautiful, until all men are wise and manifest their wisdom by a good life. The treasures of the masculine understanding and their utilization in obedience to the divine commandments, determine the state of feminine beauty in the world. Barbarians are invariably homely and repulsive, because the women have nothing to reflect in the shape of beauty, except a low, selfish, stupid and sensual type of masculine life. The higher and nobler the achievements of man over nature and himself, the more fully and radiantly does woman shine forth as the beautiful realization of his dreams. The existence of such men as Homer and Æschylus and Plato and Socrates, was the cause of the beauty of Grecian women, and of the passion of Grecian genius for a beneficent and lovely nature.

The love of the beautiful in nature is, indeed,

near akin to the chaste and spiritual love of woman; for nature reflects the wisdom of God just as woman reflects the wisdom of man. And poets and artists who project the interior wisdom of their souls into the beautiful forms of painting and statuary, music and song, exercise a kind of womanly office for the rest of men, stimulating their affections, expanding their thoughts, and developing within them the higher and nobler manhood.

This is also the secret of the fact, that the dullest imagination is kindled into some temporary appreciation of the beauties of nature by the inspiring influence of woman's love. The most stupid lovers are poets; but it is only the true poets,—born, not made,—who can detect the identity of the beautiful in the spiritual and the natural spheres.

> "She seems the life of nature's powers;
>   Her beauty is the genial thought
> Which makes the sunshine bright. The flowers,
>   But for their hint of her, were naught."

Woman's passion for dress is duly accounted for by her intuitive perception of the fact, that it is her special mission to secure the love of man by revealing to him the beautiful, although she is ignorant of the intimate relation existing between female beauty and masculine wisdom. Dress, including all that relates to the preservation and ornamentation of the person, is and ought to be a fine art with woman. There is a clothes-philosophy pregnant with more spiritual truth than Carlyle and his school ever imagined. She who does not seek to beautify herself, especially after marriage, and for the influence of beauty in the home-circle, has not fully comprehended the nature of love or the duties of life. A woman without that assistance which a refined and delicate taste can give her, is like a spring without flowers, a feast without music, a night without stars.

In an unbelieving and sensual age, woman is in danger of forgetting that physical beauty is only the lowest form of the development of the beautiful. If she loses sight of moral and intellectual beauty, and thinks that man cares only for physical charms, and dresses accordingly, she may become a vast power for evil and demoralization in the social sphere. The true wisdom of man is measured by his appreciation of spiritual more than of sensuous beauty, —or of sensuous beauty only as fitly embodying the spiritual.

The beautiful garments, the splendid jewels and precious stones, and all the graceful and charming decorations of woman in the state of highest civilization, are simply the exterior representative symbols of the spiritual treasures which exist in the heart and soul of the true wife. These treasures are chastity, meekness, modesty, fidelity,—the luminaries of the little heaven of home; the face unruffled by passion; the eye beaming with tenderness; the voice modulated by gentle and kindly emotions; the mind full of radiant and happy thoughts; the heart throbbing with self-sacrificing love of husband and children; and a soul sweet and pure, reflecting to earth the light and peace of heaven.

Alas! why is it that the proud, the sensual and the vicious can array themselves in the fine linen and purple of the angels, while the good and the pure are frequently condemned like culprits to the humblest and coarsest garb of poverty? Why is it that beautiful faces con ceal cruel and selfish hearts, while homely features fail to express the secret spiritual beauty of their possessors? The answer to these questions involves the spiritual philosophy of evil, and its effects upon the human soul and its environments,—subjects foreign to the scope of this little volume. Suffice it here to say, that by sin the correspondence between the internal and external spheres has been lost, the exterior no longer perfectly representing the interior; while the efforts of the Divine Providence are still directed to preserve an orderly and beautiful exterior, to serve as the basis of a new and higher interior life.

In heaven, however, all are young and beautiful, and clad in shining garments. This is because they are all perfect and exquisite forms of married love and wisdom. Their moral and spiritual beauty resides in a splendid external form, like a gem in its casket, like fragrance in a flower. The heavenly symmetry of their minds, the complete union of goodness and truth or of charity and faith in their affections and thoughts, gives their exterior embodiment a most resplendent comeliness corresponding to their spiritual natures. The secret of all this is the conjugial love—the vital and organizing principle of heaven, the primal source of all beauty, and the image of God in the heart.

# CHAPTER VII.

## *THE SPIRITUAL USES OF MARRIAGE.*

WE approach now the most difficult yet the most pleasing part of our task,—the spiritual uses of marriage. The subject is solemn, sublime, and inexhaustible. He who would popularize the subtile and profound teachings of Swedenborg on conjugial love, and show how marriages on earth may be made to resemble marriages in heaven, would render a greater service to the human race than the inventors of oil and wine. Many candid thinkers have recognized the immense additions which this writer has made to a rational psychology; but only his earnest students see the important bearing of his doctrines upon the conduct and duties of every-day life.

Marriage exists as a *principle* and as an *institution*.

We have shown that marriage as a principle

is universal. A principle is a general truth; a law comprehending many subordinate truths; a settled law or rule of action; a primordial element or operative cause. Nothing exists in the universe but from the union or marriage of two other things. Every form is dual. Every form is governed by laws of affinity and attraction. Sex, love, and marriage are universal and eternal.

An institution is that which is appointed, prescribed, established by authority, and intended to be permanent. The institution of marriage is a form of divine order, by means of which marriage as a principle, an eternal law or operative cause, can govern in human and angelic societies according to the divine commandments.

The institutions established by God, are designed to bring the spiritual and natural worlds into harmonious relation with Himself, so that one God, one law, one life, may govern in both, man become the image of God, and earth the image of heaven.

The written Word of God is an institution, appointed, prescribed, and established by divine authority, connecting heaven and earth. Angels live in the sphere of its spiritual life, and men in its literal sphere. Obedience to the Word in the letter elevates man from the sensual to the rational degree, and prepares him for the reception of the spiritual or angelic life.

The Church of God, with its ordinances, baptism and the holy supper, is an external institutional form by which our lives, spiritual and natural, may be brought into harmony with the Divine Will.

The marriage of one man with one woman in the external world is an institutional form, regulating the mutual influences of the sexes on each other, and turning them into specific channels; and it is an important means of attaining to the higher or spiritual marriage.

The Word of God exists in heaven, based upon its literal or institutional form on earth. The Church of God exists triumphant in heaven, based on its militant or external form on earth. Marriage exists in heaven and pours down its harmonic spiritual life into marriages on earth, just in proportion as the married partners here love the Word and the Church of

God, and lead a life of righteous endeavor and heavenly use.

Marriage was established by God; for Jesus said:

"Have ye not read, that He which made them at the beginning, made them male and female?

"And said: For this cause shall a man leave father and mother, and shall cleave to his wife: and they twain shall be one flesh.

"Wherefore they are no more twain, but one flesh. What therefore God hath joined together, let not man put asunder."

The institution of marriage has two great spheres of use, one within or above the other. In this twofold operation, first upon the external man and next upon the internal, it also resembles the two great institutions with which we have compared it, viz.: the Word and the Church.

The first great use of the institution, is to elevate man from the sensual into the rational degree of life,. preparatory to the opening of the spiritual degree in him.

Its second use is to implant the conjugial

principle in him as a means of regeneration, and so to spiritualize and fit him for eternal marriage in heaven.

Marriage is a means designed for the spiritual creation and up-building of man out of chaotic materials into the order and beauty of rational existence, and finally into the glory and felicity of angelic life.

The necessity of marriage as a civil or religious institution, which limits the intercourse of the sexes to one husband and one wife, can only be seen in the light of the spiritual truths enunciated in the preceding chapters. There are currents of spiritual life which flow from each sex to the other, and back again; currents as real and as potent as the red tide of blood which passes from the heart to the lungs, and back again, enriched with aerial treasure, to the heart.

When man and woman live together in marriage, the woman becomes so powerfully and organically impressed with the spiritual image of the man, that it sometimes happens that the children by a second husband resemble a first husband long dead, but whose spiritual image had been ineffaceably stamped on the interior forms of the woman's life. The man receives similar spiritual influx from the partner of his bosom; and his after-life is not the life of his own intellect, but the life of his intellect modified by the inflowing and inspiring emotional life of the woman.

The state of man and woman living together in the bonds of wedlock, is vastly different from the state of man and woman unmarried. The tendency of their reciprocal reactions resembling the magnetic induction, is to engraft the nature of each upon the other, and gradually to bring thought, feeling, and life into spiritual harmony.

Thus a husband or wife is a direct, powerful, concentrating focus through which the other party may receive special influx from heavenly sources.

The effect of marriage, therefore, upon the masculine soul, is to sharpen and brighten its faculties, to enlarge and humanize its powers, and to qualify it in an eminent degree for the uses and labors of life. How sweetly does the poet Schiller state this psychological fact in a letter to a friend several months after his marriage!—

"It is altogether a different existence by the side of a dear wife, from such a desolate and lonely one as mine was even in summer. Now for the first time I wholly enjoy fair nature, and live in it. Things about me again array themselves in poetic forms. What a beautiful life I lead now! My existence has gained a harmonious continuity; not passionately excited, but clearly and calmly my days flow on. A few years more, and I shall live in the full enjoyment of my mind—nay, I hope to return to my youth; an inward spring of poetic life will restore it to me."

Poetry and romance have hitherto been busied with the passionate strivings of congenial spirits after their destined union. Wedded love is yet unsung, unwritten. The poet of marriage, who shall divine the celestial secrets of womanhood, and, reverencing all while he loves only one, shall ravish mankind with his chaste melodies, has yet to appear. Not Cupid but Hymen is the true joy-bringer, miracle-worker, and heaven-revealer. From the wedding-day the heart begins a new life, the understanding creates a new history. Heaven upon earth is to young lovers a dream, but to the truly married it becomes a possibility.

Woman, being the complement of man, taken out of him and yearning to be reunited with his bosom-life, is gifted by Providence with powers of perception, attraction, submission, concealment, and endurance, and with wifely mysteries and faculties, unknown to man, whereof she is often herself unconscious, but which enable her to draw, to bend, to soften, to inspire and modify the masculine soul, so that she becomes indispensable to it, inseparable from it, and an integral and eternal part of it.

This wonderful spiritual interchange between the sexes, tending to develop a pure and true humanity, is only possible between one man and one woman in the bonds of matrimony. The wife is the true source of spiritual heat, and the husband of spiritual light, to the world. A society in which the sexes live promiscuously, has no spiritual heat or light, but only natural. They cannot rise above the barbaric standard. If it were possible to keep any free-love association together, separate from the rest of their own race, they would become in a few generations savage or imbecile.

The first or natural use of marriage, after the procreation of the species, is to elevate man above that sensual life which he shares with the brutes, into a capability of scientific and rational thought. This is below the spiritual degree of life, but is preparatory to it. The state of marriage is the key to the civil, social, and institutional life of a people. Monogamy means civilization; polygamy means barbarism. No science, no literature, no liberty, no high and progressive development, no truly rational life, is possible where woman is degraded and marriage unregulated by the laws of divine order.

The first or Asiatic civilizations drew their vitality from the golden age of the world, before the relations between the sexes became corrupted and sensualized by polygamy and concubinage. The stagnation or decline of a people dates from the desecration of marriage as an institution. The Orient has been crystallized or petrified for ages; so to remain, until some touch of divine power shall free woman from her sensual bondage, and imbue her with that spiritual love which alone can stimulate into glad activity the entire intellectual life of man.

As far back as historical records extend, polygamy and concubinage have existed in Asia; but that mankind began its career with a purer system, we have the words of the Lord Himself to prove; for Jesus said:

"Moses, because of the hardness of your hearts, suffered you to put away your wives; *but from the beginning it was not so.*"

"And I say unto you [restoring the primal, divine law], Whosoever shall put away his wife, except it be for fornication, and shall marry another, committeth adultery."

The following beautiful extract from a Hindoo epic composed two thousand years before Christ, and therefore one of the most ancient writings in the world, shows a wonderful conception of the strength and purity of a wifely love, such as existed in that "beginning" to which the Word refers. Sita, the heroine, insists upon accompanying her husband into a fearful and interminable forest into which his enemies have driven him:

> "Son of the venerable parent! hear!
> 'Tis Sita speaks. Oh, art thou not assured
> That to each being his allotted time
> And portion, as his merit, are assigned!
> And that the wife her husband's portion shares?
> Therefore with thee this forest doom I claim.
> A woman's bliss is found, not in the smile
> Of father, mother, friend, or in herself.
> Her husband is her only portion here,
> Her heaven hereafter. If thou must indeed
> Depart this day into those dreadful shades,
> I will precede and smooth the thorny way.
> Chide not the wife; for where the husband is,
> Within the palace, on the stately car,
> Or wandering in the air, in every state,
> The shadow of his feet is her abode."

While the Persians, Assyrians, and Egyptians, having closed the avenues of spiritual life, through the perversions and desecrations of marriage, were consuming away in the baleful fires of their own sensuality, two sturdy races were preparing to assume in turn the empire of the world. The Greeks and Romans were the first nations of antiquity who restored by legal enactment the lost institution of monogamy, and threw the shields of law and public opinion around the sacred rights of husband and wife. From the influx of the divine love and wisdom through the appropriate forms of the monogamic marriage, came a new institutional life—a new civilization with its monuments of wisdom and beauty, the very ashes of which still excite the admiration of mankind.

When they, too, became wholly corrupt and sensual; when religion was silent and virtue extinct; when the poet, the philosopher, and the artist disappeared, and the shadows of a long and fearful night gathered about their departed glories; any one acquainted with our key to the spiritual philosophy of history, might have foretold where the sun of a better and nobler civilization would arise, on reading a single passage in Tacitus descriptive of the German branch of the great people of Northern Europe:

"They take only one wife; and become as it were one body and one soul, having no thought or desire beyond; and loving, not only their partners, but matrimony itself."

Yes—the marriage of one husband with one wife is the keystone of the social arch. A state rises to power, glory, art, genius, with virtuous marriage. It trembles on the verge of destruction when the reciprocations between the sexes are more sensuous than spiritual.

Hence the importance of preserving the external form of the institution, as a basis upon which its spiritualizing power is to rest. Development depends upon conditions. Spiritual forces can only operate through organic forms. The institution of monogamic marriage, protected by law, cherished by public opinion, and hallowed by the religious sentiment of a people, is an organic social form into which spiritual forces can descend, and enrich our human life with numberless individual and public blessings.

It is necessary for the peace, dignity, and spirituality of the institution, that the external bond should continue in force during the period of natural life. To disturb the fixity of marriage, is to unsettle the foundations of human society, and pervert or prevent the influx of heavenly powers into the human race. It is the road back to barbarism.

The divine law that nothing but adultery shall justify divorce, is founded on the spiritual nature and uses of marriage. So long as the parties remain faithful to each other, no matter what their incompatibility of temper and character, so long does the spiritual interchange between the sexes continue, striving for equilibrium and unity: and so long is there a hope that marriage as a discipline, even through its trials, sorrows, and contentions, may promote the spiritual welfare of the parties. But when the woman absorbs the intellectual sphere of another man, or the husband becomes inspired by the love-principle emanating from another woman, the laws of our spiritual life are violated; the order of heaven is broken; the unitizing end of marriage is lost; the marriage itself is annulled; the heart is hardened; the understanding is darkened; and heaven is closed to the offending party, only to be opened again by many tears and deep repentance.

Most men are ready to admit that the institution of marriage, with all its cares, affections, and responsibilities, its love of children and home, its affinity for law and order, its kinship to religion and true liberty, has been the grand stimulus and agent of development in elevating man from a selfish and sensuous state of life to a higher and purer condition, full of thought, sentiment and rational power. And yet this high state of civilized, scientific and rational life, is not spiritual life; nor is the purest and sweetest married love ever developed under its influences, the conjugial love of Swedenborg, or the bond which unites the angels of heaven in marriage.

Spiritual life is not attainable by arts or sciences or philosophies. Men may become Bacons, Franklins, Comtes, Faradays, without having as much spiritual life awakened in their souls as may exist in some little Sunday-school scholar, or in the heart of the humblest slave. Spiritual life is only obtained *by obeying the commandments of God from religious principle.* The conjugial love is a part—and the richest, heavenliest part—of the spiritual life.

Man from sensual becomes, first rational, and then spiritual. The rational is the connecting bond by which the spiritual is enabled to descend and govern and regenerate the sensual man. The Word of God in its literal sense, and the Church of God as an external institution, have performed a mighty use in the elevation of man from his sensual state into rational light. The institution of monogamic marriage, based on the same divine principles, has exercised a similar function.

The Church, having prepared the human mind by its ceremonial laws and ordinances and literal instructions for a higher and more interior life, comes to us glorified in a more spiritual form, not yet recognized by men. The Word of God, having brought the sensual mind by its heavenly influences into rational life and order, is now opened in its spiritual degree, and all are invited to partake freely of the water of life in a higher sense than ever before. The institution of marriage, having contributed its share in the elevation of the race, is now revealed to us in new power and glory, as the great means of individual regeneration and of a perfect social organization.

Spiritual marriage is such a conjunction of minds and hearts, that the will of one party is perfectly and reciprocally united to the understanding of the other—so that the husband and wife find their whole life distinctly imaged or repeated, each in the other; constituting the "one flesh" in the symbolic language of Scripture.

Such marriages exist among angels and are possible among men, because the sexes are unequal and imperfect when apart; but, being complements of each other, they constitute, when spiritually united, one man, one angel, the image of God, and the unit in the Lord's Church.

The passional attraction, the connecting bond in the spiritual marriage, is the conjugial love, transcending all our earthly conceptions of love, as much as its delights transcend all the pleasures of the senses and all the delights of our natural life.

This conjugial love and this heavenly marriage are only possible between two regenerate souls. This at once lifts the conjugial love of Swedenborg into a sphere of thought never before traversed by poet or philosopher, from Sappho to Michelet.

Regeneration is the new birth; the birth of a new will and of a new understanding. No man and woman in the world can possibly be united in conjugial love, so long as they retain the old or carnal will and understanding. Our natural will is too selfish to blend in married harmony with the will of another. Different feelings, different opinions, discords of some kind, are inevitable between man and woman in their unregenerate state, not yet spiritualized by the new birth in the soul.

When a new will or a new selfhood from the Lord is planted in the heart, the old selfish and sensual nature dies and disappears; and the conjugial union is gradually and progressively effected from a spiritual or divine stand-point, which is the centre of all union, peace, and love.

That stand-point is attained by both parties only through *a life of obedience to the divine commandments*. These commandments are not the mere edicts of a king to his subjects. They are the organic laws of the Divine Life; they are God's own mode of existence and action. Obedience to Him is, therefore, *living like God* in our finite spheres, so that we are in perpetual communication with Him, receive life from Him, and thereby become his images—his children.

Man and woman, therefore, advance toward the state of conjugial love, and of spiritual marriage as it exists in heaven, in exact proportion as they die to their sinful and selfish nature, and rise into angelic states of married charity and faith. And the sphere of conjugial love with its delights, is perpetually flowing from heaven into married partners who strive to perform all their duties with religious fidelity; and is steadily inspiring their merely sensual and social loves with a higher and purer life.

Man and woman are not conjugially united, until the husband regards all other women from his wife's stand-point; until he has absorbed her womanly nature so thoroughly, that he feels toward all other women just as she does; nor until, on the other hand, the wife is so thoroughly impregnated with the masculine wisdom of the husband, that she sees all other men and things just as they appear in his eyes. Where such a union exists, infidelity, discord, coldness, are impossible. Where such a union exists, Swedenborg's pictures of married life in heaven do not exceed the bounds of rational belief.

This wifely struggle after identification with the husband, involving the subjugation of her own emotions and instincts, is finely described by Shakespeare in the words of the chaste and proud Queen Catharine, when defending her character before her sensual and cruel lord:

> "Heaven witness
> I have been to you a true and humble wife;
> At all times to your will conformable;
> Ever in fear to kindle your dislike;
> Yea—subject to your countenance; glad or sorry
> As I saw it inclined. When was the hour
> I ever contradicted your desire?
> Or made it not mine, too? Which of your friends
> Have I not strove to love, although I knew
> He were mine enemy? What friend of mine
> That had to him derived your anger, did I
> Continue in my liking?"

All attempts at spiritual union without a new and regenerate will, must prove utter failures. Such unions are the mere surrenders or subjugations of one will to another; or compromises of sentiment and opinion, effecting a certain external harmony and peace necessary to the well-being of both parties. The spiritual identification only takes place in the new will, the new creature. When that is perfected, the conjugial wife answers to the description given by a new poet, and which, with a change of gender, is applicable also to the conjugial husband:

"The wife is a woman married in every organ and in every fibre;
Not in body only, but also in soul.
Her eyes are married eyes,
And turn full upon one only;
From all others they are averted.
Her ears are married ears,
And hear only one voice.
Her hand is a married hand,
And clasps but one other.
Her breasts are wedded also:
All her senses are connubial.
All her life is conjugial interchange;
And in everything is the conjugial embrace."

Now comes the great and practical question,—How is the institution of marriage to be made instrumental in the regeneration of the soul, and in the preparation for that purer and eternal marriage in heaven? What special advantage has a married over an unmarried person in this respect? What is the difference between a married man, striving to obey the commandments of God, and a single man with the same religious spirit and endeavor? The answer will reveal to us the highest spiritual use of marriage.

Let us revert to, or restate, some of our fundamental psychological truths.

Sex, love, and marriage are universal and eternal; and the ideal universe is a universe perfectly married or equilibrated in its male and female elements.

The Lord infuses love or spiritual heat through the feminine form, and wisdom or spiritual light through the masculine form. Heat alone, or light alone, is powerless; combined or married, they produce all things.

Every male form in the universe has a female form, its complement, its eternal and necessary counterpart; and these forms, having specific affinities, are ever striving for union.

The Lord has instituted the marriage of one man with one woman as a means whereby the love of the sex into which we are born, shall be changed into the love of one of the sex only, and the marriage of spiritual heat and light or of love and wisdom, be effected in the soul.

The woman increases in love and communicates it to the man, and the man increases in wisdom and communicates it to the woman, exactly in proportion to their obedience to the divine commandments. Obedience to the divine law draws down the divine life, or opens the soul to its influx. The divine life is love and wisdom. The woman becomes more and more loving and beautiful; the man more and more wise and strong. Each transfers his all to the other, and thereby continually increases his own joy and perfection.

Conjugial love, thus generated in two congenial and correspondent souls, causes an intense vivification of all delights, from the highest to the lowest, from the inmost shrine of the soul to all the senses of the body.

Alas! that these beautiful and holy truths should be so little known in the church and the world!

An unmarried man receives influx into his

love-principle from the whole sphere of the female sex, which generates in him the love of the sex. However sweet, tender, and elevating, in a civilized country instinct with scientific and rational life, this general feminine sphere may be, it cannot have the direct and concentrated power of the love-element of some one woman, absorbing from him his corresponding wisdom-element, and returning it to him through undiscoverable avenues, vivified and utilized for a noble life.

When the soul fails to discover the congenial love which identifies its enraptured being with that of another, it reaches out for something to grasp, to keep, to give itself to, to blend itself with. Those least elevated above the brute, marry themselves to pleasure and money. One sighs for fame with the madness of a lover; another sees in art the ethereal bride of his aspirations. The ecstatic soul is wedded to the church, and lays his life down at her altar. The wifeless grenadier of Napoleon espouses *la belle France*, and dies in the embrace of her colors.

Marriage is an institution which brings new

influences to bear, which causes a direct and reciprocal and powerful spiritual current from one sex to the other, capable of producing incalculable evil or incalculable good. If the parties religiously keep the commandments of God, and discharge their duties toward each other with conscientious fidelity, they enjoy immense advantages over those not married. For there is a constant interchange of properties which tends continually to elevate them and unite them together. They take on each other's mental states. The woman absorbs the interior will of the man and blends it with her own which she has consecrated to the Lord; and the man elevates her understanding into that spiritual light into which his own mind has penetrated by loving and reverently obeying the Divine will. They grow more and more alike interiorly, increasing their spiritual power and perception by their union. The man rises into higher states of wisdom, the woman into higher states of love; and so, by mutual help and inspiration, they approach ever nearer and nearer to the Lord, the fountain of all love and all wisdom.

Swedenborg says that no children are born in the spiritual world; but that, instead of children, there are "spiritual prolifications" or increments in love and wisdom and the delights that flow therefrom, in proportion to the faith and love which the married parties exercise toward the Lord. The same "spiritual prolifications," or advances in the regenerate life, are obtained in this world by marriage, when it is entered upon from religious motives, is inspired with religious activities, and its duties and all the varied duties of life are performed with religious fidelity.

All the circumstances of married life—its loves, its hopes, its duties, its fears, its toils, its sorrows—tend to the destruction of our selfhood and the consecration of one's life and labor to others; to the awakening of gentle and pure affections; to lessons of patience, forbearance, self-denial; and to the cultivation of all the truly Christian graces. The uses of the married state are far better mediums for the influx of divine life and power, far better fitted to make men spiritual in the angelic sense of the word, than any celibate conditions whatever. If angels were visible, they would be seen much oftener in the mechanic's workshop and the children's nursery, than in the hermit's cell or the nun's cloister. Heaven is a state of useful, joyful labor, not for self but for others.

But suppose the parties are uncongenial and unloving, and do not tread with equal step the path of the regenerate life? Where, then, are the spiritual uses of marriage?

The Lord governs in marriage as in all things else; and we are led by ways unknown to ourselves into states best adapted to our spiritual requirements. A good man with a proud, selfish, cold-hearted woman, may gradually impart to her the grand inspiration of his spiritual wisdom, and stimulate her to a better emotional life. And on the other hand a good woman may gradually infuse a tender, womanly element into the cold, selfish, and sensual husband. And so even uncongenial partners, by striving to perform all their duties religiously, may greatly contribute to the awakening and growth of spiritual life in each other.

It may frequently be necessary that parties

wholly uncongenial, or even antagonistic, shall be brought together in this world for their own spiritual good. Deep-rooted evils which otherwise might remain unknown, may thereby be brought to light, so as to be acknowledged, repented of, and put away. The very contentions and collisions of the married life may be serviceable in the great work of regeneration, by the secret spiritual influences of the sexes on each other. The essential thing on earth is, not conjugial union,—a thing at present extremely rare,—but the preservation of the conjugial principle in the soul, which may flower after death into the conjugial love, and lead to a true marriage in heaven.

Every human being, male or female, mated or unmated, who is learning to love by striving to obey the Lord, is developing the conjugial principle in the depths of the soul, which will finally lead through every labyrinth to that glorious mansion in the skies, where the other congenial, beautiful soul awaits its coming.

A still grander use of marriage, and one running parallel with that of perfecting the individual character, is its use in the construction of a perfect society, a social millennium, the reign of Jesus Christ on earth.

Heaven is so perfectly organized that its social elements never conflict or jar; for all in heaven are regenerate, and are capable of, and are introduced into, the heavenly marriage. The same social order, peace, and love, will be approximated on earth, in exact proportion to the increase of regenerate and regenerating couples here below.

"The family," says Tholuck, "was God's first Church." And his last and eternal Church —the New Jerusalem—will be a vast congregation of families, each of which will be, through regeneration and true marriage, a perfect church, a heaven in miniature.

It is a psychological law announced by Swedenborg, that *the state of conjugial love between man and woman, determines all other loves;* determines not only the joys of the married life, but the integrity, purity, wisdom of the parental love, the love of the neighbor and one's country, the love of the Church and the Word; determines the nature of all the outflowing affections of the human spirit; determines, therefore, the entire civilization and outward condition of a people.

Give us individual regeneration; then let male and female regenerate souls be linked in holy wedlock, and we have the levers which will move and change the face of human society. The new era, the new life, the New Jerusalem, will begin in the closet, the chamber, the parlor, the home; and thence it will radiate like the sun in his glory, until it illumines and melts the world.

The spiritual wealth of one generation will be transmitted to another. Swedenborg says that sons born of parents who live in the conjugial love, are more receptive of divine truth, while the daughters are more receptive of divine love, than others. Herein lies one great hope of the world. Few men or women have the least idea yet, how far the human race may be perfected by the cultivation of maternity and paternity as high and holy arts, divested entirely of the sensual element. When natural and divine laws are studied, reverenced, and obeyed in the relations of the sexes, we shall have a strong, pure, healthy, wise, and loving generation of men, of which the highest examples we can now present are types and prophecies.

The enlightenment of the mind and the purification of the heart in both sexes, and the proper organization, protection, and sanctification of marriage, are the great forces, operating from spiritual stand-points, which are to reorganize society. Whatever social, legal, political, or religious revolutions are necessary to give these forces their legitimate field and power, will inevitably be accomplished. Institutions, governments, churches, will all be modified or perish, if they impede the advance of the great truth that the married love and wisdom of Man and Woman are ultimately to govern the world.

# CHAPTER VIII.

## *PRACTICAL TENDENCY OF OUR VIEWS.*

ALL truth has relation to life. And one of the best tests, whereby to determine the truth of any social, moral, philosophical, or religious doctrine, is its obvious practical tendency. The tendency of error is always more or less injurious; but truth ever tends to refine, exalt, and bless humanity.

If the strata of the earth are the stone-leaves of a spiritual volume; if the language of colors is a treasury of wisdom; if the voice of cataracts is a hymn of praise, and the perfume of flowers is incense to God; why should not the beautiful doctrine of conjugial love come home to our hearts as a guiding light and a continual balm? It must and will if it be, indeed, true.

Swedenborg's teachings are all eminently practical. He is called a visionary and an ideologist only by those who know little or nothing of him or his writings. His philosophy and theology are thoroughly utilitarian. USE is a term of special importance and frequent occurrence in his system. The divine life flows into angels and men in exact proportion as they delight in uses, or love to be engaged in employments useful to themselves and others. There is no place in his heaven for idlers or busybodies or recluses, or subtle theorizers who do not reduce their theories to practice. The Christian life, as he depicts it, is one of rational, genial, and constant activity.

A belief in the existence of heaven and hell, and the habit of looking forward and upward to a purely spiritual existence, has of itself an elevating influence on the soul. That power is vastly enhanced when the mind has obtained a rational insight into the laws and phenomena of the spirit-world. When we realize that heaven is near and within us, and when we form a clear conception of its organization, inhabitants, occupations, laws, etc., we have something fixed and definite wherewith to compare or contrast similar or dissimilar things on earth. We see and feel the difference between the love of self and the world which governs here below, and that supreme love of the Lord and the neighbor which gives such transcendent beauty and joy to the better life. We estimate earthly things at their true value. Theology is divested of its gloom and shorn of its needless mysteries; while death, seen in its true light, is stripped of its terror and becomes only the resurrection.

Especially useful in the daily conduct of life and battle with the world, will be the beautiful doctrine of the New Church concerning the eternity of sex, love, and marriage. Marriage as a principle and an institution is as inevitable as birth or death. What kind of association or intercourse we shall have with the other sex after death, will depend upon the marriage union of the good and the true, or of the evil and the false, in our own souls. Our life here determines not only whether we go to heaven or hell, but what scenery we shall find around us there, what house we shall live in, what companions we shall have, and the whole tone, color, and character, subjective and objective, of our being hereafter.

The doctrine that a man and woman united by conjugial love and made husband and wife, constitute spiritually one being, so that each is complementary of the other, leads to the great truth that monogamy, or the marriage of one husband with one wife, is a fundamental principle or law of the universe. All departures from it are deviations from the divine type. Each man, each woman, has but *one* complementary or necessary and co-ordinate other half. Fornication and prostitution are, therefore, deadly evils, suffocating the spiritual life and closing heaven to the soul. Celibacy is a wasted being, an unfulfilled destiny. Divorces and iterated marriages are proofs and measures of "the hardness of our hearts," and the predominance of the external over the internal life.

The doctrine that men and women joined in wedlock exercise special transforming or modifying influences upon each other, is of incalculable importance. Marry whom you will, you will never be the same that you were before. You will unconsciously absorb another life into your own, and the two currents will blend in your character and conduct. The woman appropriates and assimilates the spiritual principle of the man; and whether she reproduces it in the form of children or not, it impresses and transforms her own interior spiritual nature.

How clearly Tennyson understands it!—

——"Thou shalt lower to his level day by day,
What is fine within thee growing coarse to sympathize with clay.

"As the husband is, the wife is: thou art mated with a clown,
And the grossness of his nature will have weight to drag thee down."

So it is also with the man. The will of the woman attaches itself to his interior will, and endeavors, with inconceivable subtlety and power, to make it absolutely one with itself. Hence the purifying, spiritualizing, influence of a good and noble woman. Hence also the fearfully demoralizing, darkening, and deadening power of an evil woman over man. "There is no wickedness," said the Son of Sirach, "like the wickedness of a woman. This great psychological law is finely taught in the tragedy of Macbeth, where the fierce will of the ambitious wife seizes upon the soul of the vacillating husband, transforms it to her own type, and imbues it with her own bloody purpose.

The cause of all the woes in married and social life, is that men and women, being themselves external and sensual, are attracted to each other by external and sensual motives. Physical beauty is esteemed above spiritual beauty. External possessions are desired more than internal and heavenly treasures. Self governs all—the love of self and the world. When sensuality, selfishness, vanity, ambition, are the evil genii which bring the sexes together, and which haunt in various disguises the altar, the nuptial couch, the domestic circle, what can we expect but discord and misery; diseases of mind and body; broken vows; broken hearts; the infernal marriage of the evil and the false in the individual soul; and the awful shadows of hell projected upon earth?

Let men and women learn that marriage is the most solemn event of life, determining our spiritual growth here and our spiritual condition hereafter more than anything else. Let a pure and holy ideal of marriage be planted in the youthful mind. Let it be super-sensuous, rational, spiritual. Let the question of the heart be, not what it is to gain, but what it is to give. Let marriage be regarded as the chief means of regeneration,—as the partner and co-worker with the Church and the Word in the salvation of the soul. Let it be the occasion of self-examinations, watchings, and prayers. Then only will it become the little wicket-gate which the Pilgrim saw, and which led to the Celestial Country.

It is the part of the man to seek and woo, because it is the business of the understanding which he represents, to search, examine, discover, and organize. It is the part of the woman to determine, to refuse or accept, because it is the business of the will, which she represents, to act from a finer perception of the fitness of things than the grosser understanding possesses.

Marriage having been agreed upon, it is proper that there be a solemn betrothal, and that pledges of affection be exchanged. Swedenborg says that some weeks or months ought then to elapse before its consummation; because by betrothal the mind of one is united to that of the other, so as to effect a marriage of the spirit previous to the marriage of the body. This period of betrothal, when the two loving souls are secretly weaving the boughs of their tree of life together, and the marriage of the spirit is being effected, is a period of most tender and romantic interest. Then it is, that the magnetic fire of love, struggling to descend into the physical sphere, can impregnate a ring, a book, a flower, a kiss, with its divine life, and vibrate exquisitely to the heart,

"Striking the electric chain wherewith we are darkly bound."

Marriage, according to Swedenborg, should be consecrated by a priest or minister of religion. This is because sex belongs to the soul as well as to the body, and marriage is a divine institution. A pure and chaste love of one and one only, is near akin to the religious sentiment in man, and yearns instinctively for the recognition and blessing of a religious teacher. To withdraw the institution of marriage from the supervision and benediction of the church, and to put it on a merely civil basis, is to inflict a wound upon the spiritual interests of mankind. It is the spirit of Faust when he meditated the seduction of Margaret.

The vows are taken!—which the poet-priest of England calls "the holy vows! as sacred as the threads of life, secret as the privacies of the sanctuary, and holy as the society of angels."

Oh, that every bridal pair in the first freshness and charm of their wedded bliss, when they are all the world to each other, and life is tinted with auroral hues, could know the wonderful and momentous secret of keeping their hearts in a perpetual spring; of retaining for ever the exquisite bloom and ecstasy of their early love; of binding their souls together in eternal fidelity, purity, and peace; and of making their home such a heaven, that heaven must inevitably be their home!

This secret is not a matter of curious speculation or of religious sentiment, but one of scientific precision. It is as fixed and certain as the theorems of geometry or the laws of optics. The love which unites, purifies, and spiritualizes, and in which there is innocence and peace, comes only from God; and woman in the act of obeying the divine commandments, is the medium of its reception. The wisdom which illumines, strengthens, and guides, comes only from God; and man in the act of obeying the divine commandments, is the medium of its reception. This love and wisdom, or spiritual heat and light, are communicated from each sex to the other, and are married in the hearts of congenial souls obedient to the Lord, so that their winter is turned into spring and their earth into heaven; for heaven is opened to their inmost souls, and its glorious light and peace and joy descend, producing all the graces and virtues of life, the happy mind and the contented heart.

When husband and wife profess and enjoy the same religion, and are equally impressed with a sense of their obligations to the Lord and to each other; when they make the Word of God their rule of life, and read it and pray over it together; when the spiritual welfare of their children is the first and united aspiration of their hearts; we have a home of heavenly order, purity, and peace, a true centre of divine life, the ideal of Christian civilization, and the hope of the world.

But when no altar to God is erected in the house or in the heart, there is no direct channel for the descent of the Divine love and wisdom. The motives are external and worldly, the love sensuous, the life animal, or at most rational, but not spiritual. The old, unregenerate, selfish, and sensual will rules in both parties. There is no true and lasting bond of union. There is no genuine Christian love of children or of each other. Everything is at the mercy of time and accident and fortune.

Such marriages, however, are not entirely without their spiritual use; for the uses of marriage are so holy and heavenly, that something of its great power and glory will flow down into its merely external form. The experiences of married life, the mutual cares, labors, responsibilities, and trials, even its collisions and temptations, are means of bringing to light the hitherto concealed evils of the parties, convicting of sin, awakening sympathy, and kindling the sparks of their better and purer nature.

The proud man who would kill his neighbor for telling him his faults, will bear the reproaches of his wife. The selfish woman whom nothing softened before, will find her heart melting down in the sweet light of her children's eyes. And many an irreligious couple who began their married life far apart in spirit, descend to the river of death, weary and worn, wounded and broken in the battle of life, but with new and enlarged sympathies, with new humility in their souls, looking upward to God.

A wise and thoughtful man yearning for the spiritual life—what shall he do when he finds himself mated with a vain, silly, selfish, and heartless woman? His wisdom seems like the light of winter, cold and fruitless; but it may accomplish much for the spiritual welfare of his household by faithfully living out, in word and deed, the duties of man to woman, of the understanding to the will, so well described by Jeremy Taylor.

"The dominion of a man," says he, "over his wife, is no other than as the soul rules the body; for which it takes a mighty care, and uses it with a delicate tenderness; and cares for it in all contingencies, and watches to keep it from all evils; and studies to make for it fair provisions, and very often is led by its inclinations and desires; and does never contradict its appetites but when they are evil, and then else not without some trouble and sorrow; and its government comes only to this, that it furnishes the body with light and understanding. The soul governs because the body cannot else be happy; but the government is no other than provision."

The pure, spiritually-minded woman, who is wedded, or rather *yoked*, to an unworthy husband who by a false and evil life is gradually effacing the stamp of humanity from his soul—what shall she do? As man works by wisdom or light, so woman works by love or heat. She fights, conquers, lives, by love. The love of a true Christian wife is like the charity described by Paul, and which, he says, is greater than faith or hope; which suffers long and is kind; which doth not behave itself unseemly; seeketh not her own; is not easily provoked; thinketh no evil; which beareth all things, believeth all things, hopeth all things, endureth all things.

This charity, in the heart of a good wife, is more powerful than "the tongues of men and angels."

A woman attaches her husband to her by attaching herself to her husband. Let her ever reveal to him the beautiful, whose soul is the good and the true. Let her learn the deep significance of a rose in her hair; a flower on the mantel; a smiling face at the door; the snatch of an old, sweet song at her work. Let her see her husband in all things; and so thoroughly identify herself with him, that he shall find the angel of his better nature in every lineament of her face and in every gentle tone of her voice.

The regenerating influences of husband upon wife and of wife upon husband are so great, so subtile, so lasting, that parties are brought together by the Divine Providence for the sake of these secret ministries of marriage. Only the revealing light of another world can show how much spiritual good has been accomplished to both parties, even where their external union seemed singularly uncongenial and unhappy. The only condition for the reception of every blessing which the Heavenly Father can give us consistently with our eternal good, is the faithful and religious performance of each one's duty.

Therefore the bonds of marriage are not to be lightly broken, and the soul permitted to go ranging after new affinities. The true spiritual wife is *surely* found only in the spiritual world. The essential thing in this world is, not to *find the conjugial partner*, but to acquire and *preserve the conjugial principle* in the heart. This may frequently be best accomplished by living with an uncongenial person, faithfully and religiously performing all the duties required by the spirit and letter of the bond, just as if he or she were "the one beautiful soul" of our celestial future.

If sex were purely physical, if marriage were only a civil alliance, if our material life were all, then our philosophy and theology would be vain. But if our souls are male and female; if marriage is spiritual and eternal; if this life is the seed-field in which the germ of a better life is planted; if wedlock is a divinely appointed means of bringing the spiritual influence of the sexes on each other into orderly and beneficent activity; if the character of the husband and wife determines that of the father and mother, the neighbor and the citizen; if a life of obedience to God in the state of matrimony is peculiarly rich in spiritual blessings; then are the teachings of Swedenborg on these lofty themes of immense practical importance in the regeneration of the individual and the reorganization of human society.

The beginning of evil as wonderfully described in the spiritual sense of the third chapter of Genesis, was the loss of married harmony between the will and the understanding. That perversion of the good and the true in the individual soul, was the cause of all the corresponding perversions in the sphere of our sexual life. The way back to that blissful state represented by the Garden of Eden, is through individual regeneration, and through marriage made pure and holy, by a life according to the Divine commandments. Swedenborg has given us some interesting accounts of marriage in heaven—accounts so beautiful, so heavenly, that our poor earth-stained souls regard them as fancies or dreams—as a grand ideal for our own aspirations, our hopes, our prayers.

When the lost harmony between the will and understanding is restored, moral and physical disorders will cease. The evil, the false, the unbeautiful, will disappear from man and nature. History will then be the harmonic development of a loving brotherhood under the smile of the Divine Father. The New Jerusalem will descend from God out of heaven, "prepared as a bride adorned for her husband." And thenceforth the marriages of men on earth will be similar to those of the angels in heaven.



FINIS

Made in the USA
Monee, IL
04 November 2026

67385011R00157